AMAZING GRACE OF QUANTUM PHYSICS

AMAZING GRACE OF QUANTUM PHYSICS

Dillard W. Faries

PICKWICK *Publications* • Eugene, Oregon

AMAZING GRACE OF QUANTUM PHYSICS

Pickwick Publications
An Imprint of Wipf and Stock Publishers
199 W. 8th Ave., Suite 3
Eugene, OR 97401

www.wipfandstock.com

PAPERBACK ISBN: 978-1-5326-1421-7
HARDCOVER ISBN: 978-1-5326-1423-1
EBOOK ISBN: 978-1-5326-1422-4

Cataloguing-in-Publication data:

Names: Faries, Dillard W.

Title: Amazing grace of quantum physics / Dillard W. Faries.

Description: Eugene, OR: Pickwick Publications, 2017 | Includes bibliographical references and index.

Identifiers: ISBN 978-1-5326-1421-7 (paperback) | ISBN 978-1-5326-1423-1 (hardcover) | ISBN 978-1-5326-1422-4 (ebook)

Subjects: LCSH: Quantum theory—Religious aspects—Christianity | Religion and science

Classification: BL240.3 F174 2017 (paperback) | BL240.3 (ebook)

10/30/17

To the memory of
Gerald Hawthorne
a beloved professor of Greek and encouraging colleague

Luci Shaw, in a poem called "'The foolishness of God' for Gerald Hawthorne," (*Polishing the Petoskey Stone,* 198) caught the essence of what I thought was Jerry's central message: God is not reasonable; he asks the impossible, but he provides the way. This message resonates with what I hope this book says, ever so weakly. Listen hard.

Contents

Preface

ANTON CHEKHOV SAID THERE are vast fields between "There is no God" and "There is a God." We have a nearly impossible task to fill in who or what God is. Those vast fields become dense forests or deep chasms. Schiller (in one of Niels Bohr's favorite quotes) echoes Democritus: the truth lies in the abyss. No wonder that faith is needed and is also so easily scorned.

But there are plenty of fields, forests, and abysses between "There is no electron" and "There is an electron." At the peak of our arrogance, we may have thought that only two numbers—mass and charge—constituted our "discovery" of this God-given fundamental constituent of Nature. As its properties became more nuanced and muddied, we can legitimately ask if we *invented* the electron or at least discovered only what our pre-conditioned concepts would allow. There may be differences in degree but the existence of the electron is epistemologically in the same category as the existence of the mythical gods of ancient Greece or the Judeo-Christian God.

Reality is slippery business. When my pre-school grandson makes a symmetrical patterns of Legos™ with a mirror image and announces the "real" dog on the image side, I know that "reality" can depend on our sorting. When a Muslim finds that quantum mechanics claims a system "really" exists in three resonances at once, it may shatter the trinitarian problem of one God. (See Qureshi's book, *Seeking Allah, Finding Jesus.*)

I was already a physics professor who had been indoctrinated and puzzled by quantum mechanics when a young woman wrote me a letter about her faith struggles upon reading a popular book relating modern physics and Eastern religions. I have not been able to shake or to solve those problems ever since. Years of this struggle have resulted in this book, a very incomplete and meager attempt. Religious questions are called out-of-bounds in science, and science is not often welcomed or understood in religious circles. These two areas cannot be held apart without damage to both, and they do not have only an adversarial relationship. We must keep trying to let them speak to each other.

Acknowledgements

First, I must acknowledge you, the readers, who are the only ones who can complete this attempted communication. There is no objective message of truth out there independent of you the subject. Authorial intent may be considered a prime hermeneutical principle, but reader's intent is just as important. Your questions, your directed attention, will make it what it is. My words are empty until they're planted. Thank you for your efforts.

I have appreciated and profited by the immense literature and the attempts to preserve and digest it. Growing up in small towns without libraries, I have relished the opportunities of libraries and archives, ranging from the small-town public library to the university mega-library systems and international archival materials now available on-line. Scientific literature, as published, is almost completely sterilized, with the notion that passive-voice depersonalized language carries objective truth. A few days in Berkeley's History of Science collection and Yale's archival materials give information sparingly but bring people to life. Published correspondence (sometimes voluminous) does the same.

My teachers, colleagues, and students have provided inspiration and encouragement and not only by positive examples. The isolation of our disciplines and the teaching of science as gospel truth of hard facts represent something to buck against, even—or especially—when we find it in ourselves. I am especially grateful to Wheaton College, its faculty, and student body for their openness to truth wherever it is found. Because of its strong faith base, it can have advantages over the clearly great universities, Rice and Berkeley, which have contributed most to my professional training and for which I am most grateful. In a similar vein, the American Scientific Affiliation, an organization of Christians in science, has been an important influence in my life. Because of the large number of individuals in many places, I will forego giving names, but I think you know who you are. Thank you for your part in my life.

I must break my rule and mention some names of influential writers: John Polkinghorne, Stanley Jaki, and William Pollard. Each of them were trained in physics and then went into the priesthood. Michael Polanyi is a similar case. And I must mention the Gifford lectures, a tremendous resource of a yearly lecture series in major Scottish universities, almost invariably published in book form. (Three of the four names mentioned here participated in this series and will be represented in the bibliography.)

Unless otherwise noted, all figures were produced in Microsoft Excel or from Clip Art in Microsoft Word. Any limited quality is due to my own inadequacies and poor choices.

Finally, I must thank my family for providing the source, the support, the time, the patience, and the inspiration to do what I can. This includes my forebears, my extended family, my children and grandchildren, and most importantly my wife Sally, *sine qua non*. Love and thanks to you all.

Introduction

> Chance is Nature's free will; choice is mankind's free will; grace is God's free will.

It is said that the most obvious empirical fact of our lives is the presence of sin. This can be seen in any stretch of history. A look at twentieth-century history easily shows the lie of civilization's advance and obviously belies the ancient Greek philosophers' belief that knowing to do good and right automatically leads to doing good and right.[1] It is not lack of knowledge that results in sin. Sin or at least its awful results are painfully visible. But sin as a personal problem, *my* personal problem, does not seem to be as obvious. Nor is the obvious root *and* cure for sin immediately seen or universally recognized. Pause and give your answer before you read mine.

I believe that the root cause and the final and complete cure of sin lie in one single word: choice. Believing, as I do, in the sovereignty of God over everything, I have to blame God and give him the credit for both ends of this vicious stick under which we live. How can *choice* be both cause and cure of our universal ailment? God, making us in his image, gave us the major characteristic of a sovereign: autonomy, self-rule or *choice*. If he does not give his creation choice, his creation is nothing more or less than an automaton, a robot, a helpless completely programmed computer, a very lawful *sinless* being, and not very interesting. God *chose* to make his world capable of sin. If you *have* a choice, Murphy saw that you *would make wrong* choices, at least sometimes. Murphy's law of the universe says: if mankind can sin, mankind will sin. If mankind was not given the choice, mankind would not sin. God took the chance, was willing to live with that chance which caused him and us so much suffering, for the opportunity to draw some of us out of that world of bad decisions.

1. Herman, *Cave and Light*, 23. Herman describes Plato's legacy as a belief that knowledge is the prerequisite of virtue, with ignorance leading to evil.

It was God's choice and our choices which put us here in one hell of a mess. (I at least try to use the h-word appropriately.) Chance and choice are two sides of one coin. If we are outsiders, not privy to the choice decision, we call it chance; if we are the insider, we call it choice. Chance means someone else is inscrutably making a choice. Because we can't see the decision process or causal chain, we, in our ignorance attribute it to chance, an uncaused happening. If we think we have figured out the system, we call it causal and make our knowledge into laws. If we cannot figure out the system, we may call it lawless. Those damned people simply made bad decisions. (Again, a literal use of a word we should use cautiously.)

And it is God's choice and our choices that can get us out of this mess. There is another word for God's choice which we Christians believe will be the ultimate winner: *grace*. We sometimes try to describe grace as *unmerited favor*.[2] *Favor* is someone else's choice that is good for us; *unmerited* means that it was unpredictable and uncaused. In the scientific language which does not admit a personal involvement and cannot see the person, let alone the internal decision-making of the person, this might be called totally random chance. Chance, choice, and grace are inextricably tied together. Chance, a truly open future as seen from our limited present circumstances, is necessary for choice. Chance is the freedom in free will, and choice is the will. Volition and will, the presentation of options and the deciding between options, constitute a coupled prerequisite, perhaps *the* prerequisite for personhood.

> 'Twas grace that brought me safe thus far,
> And grace will lead me on. (John Newton's "Amazing Grace")
> 'Twas choice that got me in this mess,
> And choice will get me out.

The problem choice was my choice, our choice, creation's choice, allowed by God in order to make a universe of meaning; the second choice is God's choice, requiring also our choice. The Reformed Mafia may insist it is all God and all law and no chance and they may have spawned that amazingly successful science, the greatest revolution since a baby boy was born in Bethlehem, a science of perfect laws just waiting to be uncovered, a science which may be discovered by reason, a complete knowledge which will surely bring in the millennium, no personal God required. It was that vaunted science which in turn spawned Marxism, communism,

2. Lewis says: "It is not our right but God's free bounty. An English peer said, 'I like the order of the Garter because it has no dam' nonsense about *merit*.' Nor has Grace." Lewis, *Yours, Jack*, 180.

and modern authoritarianism as political copies of the presumed physical world.[3] And, oh yes, wait! The great American political system was spun out of the same historical cloth.[4] And the economic theories which grew from the Newtonian fountainhead included both the Marxist socialist world and our free enterprise system. With all these connections, an outsider might see Marxism and Christianity as twin religions, Marxism taking the material form while Christianity sometimes tries to become totally spiritualized.

Fortunately, physics in its classical Newtonian form, which spawned an enlightened, rationalistic, optimistic, we-can-know-and-do-it-all attitude, ran into a dead end, even within its own self-imposed materialistic ghetto. We know that the materialist viewpoint of Carl Sagan, that the Cosmos is, was, and will be all there is, cast into a utopian pure and perfect and complete knowledge is a false idol; it never could work. While the textbook pictures and public image of science remain true to that absolutely false Newtonian pseudo-ideal, physics has been forced to build on a new foundation, a foundation much more in tune with the fundamental Christian and Jewish and Muslim message of a personal God who came and comes into the world in all kinds of manifestations: a creation, a chosen people, a set of laws, a place of worship, an incarnation, a resurrection, and a host of prophets who speak his word. The most revolutionary aspects of modern physics are found in the world of quantum mechanics (QM). Thus I have been privileged to explore the wonderful message of God's grace as seen in the last century's developments in modern physics. If you like, it's the "gospel according to QM." Those raised under the thumb of the evangelical dictatorship may not believe that anything outside the Bible can reveal the truth. Our communications, our very languages, spring from our physical experience in the world. The Bible cannot be read literally, it literally cannot be read, aside from how we conceive and express the physical world around us. When the physical world is reconceived, we will over time have to learn anew how to read the Bible and state our beliefs about God. I am way too limited in almost every way to make any small dent in this enormous project for our age. I am only thankful for the small insights I am able to imperfectly glean. I hope you will find my sharing useful, not as answers, but as a stimulus for further study and discussion.

Our culture, our worldview, is the water we swim in, absolutely invisible to us because we were fed it from our mother's breast, we grew up in

3. See, for example, Mumford, *My Works and Days*, 354. "The orthodox communist has not escaped the mechanistic prison."

4. No less a personage than Woodrow Wilson is quoted as saying: "The government of the United States was constructed upon . . . a sort of unconscious copy of the Newtonian theory of the universe." Rukeyser, *Willard Gibbs*, 79.

it, and we assume there is no other world or way. When you really want to know about the world, when you dare to approach the difficult questions, you eventually will need to question some of these givens. Not everybody at all times sees with the same eyes that you see with. We as Christians are accused of making God in our image. Whether our concept of God is our pure imagination which is limited to what we know, or God making us in his image, or God communicating to us in the only language we could possibly have, we may never know. But this is not only a theological problem. Everything we believe, absolutely everything, has the same problem. We have fundamental beliefs which affect everything we are capable of analyzing. Most of the time we are incapable of analyzing the fundamental beliefs because they are invisible. They may hold the whole structure up, may make the whole structure possible, may shape the whole structure, and may be the creator of the whole structure. But we, most of us most of the time, are totally unaware they exist. And who can know? They may be all right or all wrong or some strange mix of right and wrong. Presumably the greatest thinkers can touch, see, shape, and use these fundamental beliefs, what might be called definitions, axioms, tenets, or self-evident claims. I can make no claims to begin to understand either the foundations or how they came about, but I hope to stimulate your thinking, as mine has been, by the challenges which the world of physics has faced and is facing. Historically the world of mathematics has often held some primary position as a fountain of certain knowledge. Geometry, by its very name "measuring the earth," seems to be tied to the material world. On the other hand, the ability to produce an abstracted, ideal axiomatic system of geometry tightly deduced from supposedly self-evident starting points raised geometry to the *sine qua non* of Plato's Academy and made it appear to *rule* the material world, to force the physical world to follow its mold, as well as an evil, second-rate reality such as matter and flesh made of matter could. Reality for Plato *was* that ideal and it was trapped in matter. We marvel that a physical world follows mathematics, the "*unreasonable* effectiveness of mathematics"[5] in a physical world, we say, as though mathematics is a free-standing tower of Truth which is independent of the Truth of the physical reality we are stuck with. But now, we have constructed a tower, buttressed on one side by the deductive world of pure mathematics which may still be able to build castles in thin air and on the other side by the inductive, real kickable world of hard data stuck deeply in the bedrock of commonsense reasonable and measurable quantities. This is the certain knowledge, the religion of modern mankind fashioned from the seventeenth-century marriage of mathematics

5. Wigner, *Symmetries and Reflections*, 222–37.

and science by Kepler, Galileo, Bacon, Descartes, Newton, Leibniz, et al. The physical world held the feet of mathematics to the fire and mathematics lifted physics to the ethereal heights. When David Hume could apply his skepticism to miracles, which he claimed was the root of religion, and equally to causality, the root of science, Kant roused himself to admit mankind's mind as the central source of the categories such as space, time, and causality which were necessary for "pure reason" and to admit the need for the "practical reason" of the religion to morally guide the actions of lesser minds.[6] Sometimes it seems that Kant rescued science, but failed to rescue religion, particularly the miraculous basis. Although Kant took his major place in modern philosophy, the categories that he saw as necessary to scientific knowledge have not stood so well. The modern scientific tower to heaven, the modern Tower of Babel, ran into serious problems with Nature questioning our overly simple views of space, time, and causality.[7] The crisis in physics and our attempted solution of it gives us a great occasion to question the invisible medium we live in. As David Foster Wallace so poignantly asks the fish's question ("what is water?") in his 2005 commencement address at Kenyon College,[8] we need desperately to ask what is the medium we swim in, a medium so familiar that it is invisible to us. Is this an opportunity for people of faith to reconceptualize God, a heretical thought to a huge portion of Christendom? Or is it an opportunity to see our concepts of God play a deciding role in what science should believe? That is, can science reconceptualize its god(s?) based on a body of faith, a suggestion equally heretical to a large portion of the scientific hegemony? In any case, a poor man's account of some of these issues may hopefully inspire you to do better than I. Good luck, Godspeed, or whatever messages of encouragement your belief system will tolerate.

The following general outline is a roughly chronological account of quantum ideas with intertwined applications to possible theological developments. These chapter summaries are too brief, but may provide a guide for the overall plan.

1. Paradoxical issues

   In this chapter, I give a *conceptual* outline, an annotated list of conceptual pairs or triplets, sometimes contradictory, sometimes complementary,

6. Raised in a religious home and taking philosophy seriously as a guide to right living, the great philosopher Kant felt no need for the religion of his family upbringing, but hoped his servant would be religious, so he wouldn't run away with the silverware.

7. Margenau says that Kant's philosophy was "the metaphysical distillate of Newtonian mechanics." Margenau, *Open Vistas*, 130.

8. Wallace, *This is Water.*

but always central to the collision of a classical world of modern science (the results of seventeenth-century physics) and the non-classical world of modern physics, especially quantum physics. Because they are inextricably mixed in both historical and conceptual development, this cannot be a neat categorical outline for orderly presentation. However, the presence of each lurks as a shadow in the background. The list includes: continuity/discontinuity, certainty/uncertainty, determinism/indeterminism, reality/a-reality/anti-reality, object/subject, causality/a-causality/accident/randomness, chance/choice, reversibility/irreversibility, being/becoming, history as progress/history as decay, relativism/absolutism, narrative/metanarrative, single unitary description/complementary descriptions, locality/nonlocality, reduction/wholeness, and causality/purpose. Some major topics, even whole chapters, manage to sneak into these categories implicitly and some of these categories seem to permeate the whole structure of the revolutionary shift.

2. Seeing beyond Seeing

   Already five years before Planck sowed the first seeds of QM in 1900, the features of chance and discrete events characteristic of the unseen quantum world burst upon us in the world of X-rays and radioactivity. This almost takes the form of a prophetic John the Baptist preparing the way for the Messiah. Atomism had been a glimmering prevision dimly seen for millennia and made useful in the world of chemistry for a century, but now the individual pieces and individual events with tremendous jumps gave a picture of the tremendous power and freedom which were possible in the unseen realm. Chance and probability were part of our lives and therefore part of our games long before they became part of our science. Statistics had been useful in dealing with partial knowledge, but now the absolute paucity of our knowledge opened up new realms of ignorance which did not appear to be conquerable.

3. God, Classical Physics, and Modern Science

   The bulwark of classical physics took root in the Christian West, intimately tied to many theological concepts. Kepler and Galileo, as Protestant and Catholic respectively, each had deep religious roots and forged a synthesis of the celestial and the earthy, the ideals of mathematical purity with religious symbolism and real-world data, the *a priori* results of a Cartesian deductive mind and the *a posteriori* Baconian induction, and the Platonic otherworldly geometric ideals and the Aristotelian organismic all-too-this-worldly physics. Newton, with a mathematical calculus of unlimited precision, fashioned the complete and perfectly accurate seventeenth-century theory of everything, matter in motion, the complete choreography of the universe.

4. This Little Mine of Light

   Light is a key, perhaps *the* key physical phenomenon and a key religious and philosophical touchstone. The history of light is an illustrious look at the history of ideas, intertwining our conceptual frameworks. The seventeenth century saw three major systems emerge in the theory of light: the corpuscular (particle) theory of Newton, the wave theory of Huygens, and the least time theory of Fermat. The least time theory was and remains an attractive idea, but the Newtonian hegemony in mechanics carried the particle theory on its coattails for the following century until a nineteenth-century revival of wave theory.

5. What Is Reality? Masses or Fields? Ether or Vacuum? Mathematics or Material World? Material World or Spiritual World?

   The world of mechanics which Newton developed gave a complete description for the motion of masses. It gave strict definition to *force* which was otherwise a vague, unconvincing concept in the case of gravitation where forces acted at a distance. Newton was quite aware of the conceptual gap covered over by an apparently perfect *descriptive* mathematical formula. When electricity and magnetism followed similar descriptive formulas but gave no better idea of how it worked, Faraday visualized the concept of fields as an intermediary to transmit forces through space. Maxwell gave the fields a mathematical formulation uniting electricity, magnetism, and optics into a set of equations which gave light a beautiful, convincing, and apparently complete wave theory. Classical physics now had two intertwined concepts, the discontinuous world of particles with mass (fundamental particles, atoms, molecules, and big objects like balls and planets) and a continuous space-filling world of fields which could propagate as waves.

6. Turning the Heat on a God Who Was Too Big

   Classical physics had two major fields: mechanics (discontinuous matter (particles) moving in continuous space-time) and electromagnetic theory (continuous electric and magnetic fields providing forces at a distance between charged particles). A third leg, called classical thermodynamics, focused on systems with very general ways of dealing with the laws independently of the unseen interiors. Although mechanics and atomism threatened to subsume thermodynamics into a statistical mechanics for material systems, Planck tried to deal with universals by looking at the thermodynamics of the continuous fields.

To his surprise, electromagnetic fields, the very successful Maxwellian complement to the presumed discontinuous world of atoms and molecules, showed a discontinuity of its own. Energies of electromagnetic modes of vibration appeared in only discrete values, like a ladder or staircase with no place to stop between rungs or steps. A new fundamental constant of Nature, called Planck's constant, h, appeared, and began the revolution which was eventually called QM.

7. Who Will Hear the Still Small Voice?

   Planck, the solid German scientist, had discovered the very small, very key universal constant, h, but it fell to a trio of Jewish youngsters, destined to become good friends, to hear that still small voice and amplify it into a major influence in the world of physics. Paul Ehrenfest, a disciple of Ludwig Boltzmann and thus a key figure in the atomic hypothesis and the reduction of thermodynamics to statistical mechanics, had a keen nose for the fundamental questions, a critical eye for subtle points, imaginative communication which transcended language barriers, and an attractive personal style which drew scientists together. Niels Bohr, a disciple of Ernest Rutherford, transformed the impossible classical model of an atom as an electron-nucleus planet-solar model, the impossible mismatch of such a model with electromagnetism, and the impossible massive array of spectroscopic data into a crude quantum model which allowed a meaningful modeling of the periodic table. Einstein, long busy with his revolutionary relativity and frustrated (though awed)[9] by the messy unprincipled style of Bohr's theory, again provided confirmation and encouragement for the quantum world by tying his belief in the particle of light to a model of absorption and emission which linked Bohr's untidy work to Planck's radiation distribution function, giving aid and comfort to what must have seemed like the enemy.

8. Improbable Person and Improbable Method Lead to Crazy New Worlds of Quantum Statistics

   Particles are somehow obvious things of brute existence, separable from the rest of existence, taking up space which is thus unavailable to other things. And the universe is divisible into particles according to our will or whim. If we can slice up space, we can at least conceptually divide the universe into pieces. Natural divisions such as planets and man-made divisions such as cannon-balls made *divide and conquer* a

9. Einstein called Bohr's work "an enormous achievement" because it was so effective in describing Nature even though it did not hold to any clear principles. To him, it was more art or music or poetry than science. Klein, *Ehrenfest*, 278.

reasonable tactic for mechanics. The inner world, whatever the smallest pieces might be, had proved more difficult and had been probed by a supposedly more comprehensive program of thermodynamics which was hopefully independent of the details. As atoms began to make themselves known, a statistical mechanics began to take over the realm of thermodynamics, a statistics of large numbers of Maxwell-Boltzmann particles, commonsense distinguishable pieces. But the world appeared to be made of two different kinds of "things": commonsense distinguishable material pieces and fields which carried energy, had momentum, moved in space and time, filled all space, but shared it with others, fields and particles. These two things should probably have different statistical characteristics, one involving the counting of discrete discontinuous individuals and one involving a presumably infinitely divisible continuous world.

Planck had quite intentionally dealt with a world of fields, a box of electromagnetic radiation. Not believing in *particles* of electromagnetic fields, he had nevertheless reached a statistical argument involving a discontinuous world, something like particles. Einstein came to believe that light came in pieces and successfully used this idea to explain how light kicked electrons out of materials. He also worried about the statistics and found that statistical fluctuations involving light had two parts, one like particles and one like fields. Ehrenfest had shown him that a particle derivation of Planck's law involved features quite unlike those of the Maxwell-Boltzmann derivations.

When an outsider, a Bengali named Bose, came up with a derivation of the correct Planck law using some questionable or even incorrect steps, Einstein saw that the particles of light must be quite different from Maxwell-Boltzmann's distinguishable permanent massive particles who hogged their own space and identity. These new particles could be created and annihilated; they shared space, and were like identical twins, triplets, quadruplets . . . , indistinguishable and interchangeable. These kinds of particles mimicked the features of fields. Not only light but particles of mass might behave this way and Einstein worked out their behavior, a weird possibility being a condensation into a single state. Such particles came to be called bosons and such a condensation is called Bose-Einstein condensation. It took seventy years before such a gas of bosons was thus experimentally condensed.

In the same time frame, still before a full-fledged quantum theory appeared, another type of particle, fermions which include electrons, protons, and neutrons, became understood as indistinguishable particles which mimicked the space-hogging exclusivity of commonsense particles. Rather than refusing to share the same space, they could not share the same quantum mechanical *state.*

The indistinguishability within each category of bosons and fermions and the obvious result that an exchange of any such pair, anywhere in the universe at any distance, must be undetectable means that the universe has a unity, a wholeness, a universe-wide connection, a super unbreakable and mysterious action at a distance, an invisible web connecting everything.

9. *Judenphysik* to *Knabenphysik*

Youth, new blood was necessary for a revolution to appear, and three baby boys were born within three years of Planck's very quiet, very uncertain introduction of the quantum world's universal constant h. Werner Heisenberg, the enigmatic Great White Hope of Germany, was overshadowed in school by his friend Wolfgang Pauli, the Viennese Jew and noted critic and superseded later by the "strangest man,"[10] the English engineer Paul Dirac. (For those who fear the label *engineer* might be a put-down, Dirac was trained as an engineer but later held the prestigious Lucasian chair held by Isaac Newton and Stephen Hawking and a host of greats in between.) But Heisenberg, armed with the optimism learned from Sommerfeld, the premier physics teacher of the twentieth century, the mathematics learned at Göttingen, the mecca of mathematics at the time, the physics learned from Bohr, the acknowledged pope of the quantum world,[11] the positivism he thought he learned from Einstein, the critical example of his good friend Pauli, the nerve of youth, and the humility borne at the hands of Wilhelm Wien, Heisenberg grabbed the brass ring, being the first to produce a formal quantum theory. It was expressed in matrix algebra, mathematics unknown to him and practically unknown in the physics community of his time, and became known as matrix mechanics.

10. *Another* Form of QM and the Loss of Determinism

10. Farmelo, *Strangest Man.*

11. Bohr was unquestionably the leader in the developments of quantum physics, often given such honor as the title of pope. In a well-known production of a take-off on Faust in Copenhagen, Bohr was portrayed as God. (Segrè, *Faust in Copenhagen.*) He received that kind of respect, often considered *the* person a young aspiring physicist must meet. Wheeler thought that the titles which suggested authority missed the mark, because the society at Bohr's institute was more like a family, including the mother Margrethe and Bohr's children. Oppenheimer states Bohr's position well: "from first to last the deeply creative and subtle and critical spirit of Niels Bohr guided, restrained, deepened and finally transmuted the enterprise." Oppenheimer, *Atoms and Void*, 29.

Heisenberg, well-prepared, well-connected, and very clever, gained priority in producing a formal mathematical theory of the quantum world which Born dubbed QM in the series of papers following Heisenberg's. I cannot imagine anyone remembering Heisenberg's original work with its generally inaccessible mathematics and its quite intentional deviation from the concept of causal motion of particles, the mainstay of classical mechanics which simply could not be followed at the atomic level. It could probably not have succeeded without the equally unpictureable concept of the wave nature of material particles and the much less abstract mathematics which was used. De Broglie completed the developing duality of light as particle/wave by suggesting that the commonsense particles of matter acted like waves just as the by-then-commonsense electromagnetic waves followed Einstein's belief in their particle nature. Schrödinger wrote a wave equation for this new wave, trying hard to attribute reality reasonably reserved for particles to his waves. He quickly solved a series of problems which still constitute a first course in QM. The logical title of *wave mechanics* may still be found in old books or in historical chapters. QM quickly absorbed a broader conceptual range and was formalized by a third young man, Paul Dirac.

11. What Is This Thing Called *QM?*

We make an attempt to get at some of the essence of both our faith and QM. The double reality of divinity and humanity in Christ is certainly the core of Christianity, and a double reality of wave and particle as the fundamental constituency of the universe is at the core of QM. Jesus *fully* divine and *fully* man? Something just doesn't add up. The outworking of this duality of Nature in the question of how things add is the essential mystery of QM.

12. I'm Not Sure!

The Heisenberg uncertainty principle is a key linchpin in the interpretation of QM. Classical waves in space or in time classically exhibit uncertainty principles. In space, it is what limits the optical resolution of optical devices. In time, the frequency determination is limited (inversely) by the time duration of the signal. If particles behave as waves, then these purely wave principles carry over into the particle world of energy and momentum with an obvious appearance of h, the constant relating momentum and wavelength, energy and frequency, in the Planck-Einstein-de Broglie relationships. These limitations of our knowledge seem to come from an ontological limitation and even seem to force randomness on us. The classical picture of sure and certain and precise knowledge is no longer tenable to the world of physics. We in the Christian faith have followed classical physics into a world of clarity and lost the mystery and uncertainty and faith which science has recovered in modern physics.

13. Complementarity: Either . . . Or . . . , Nothing But . . . , Both . . . And . . .

Bohr saw complementarity, the necessity of two apparently contradictory concepts or viewpoints, as being a key idea in the interpretation of QM. Although he saw this as an epistemological lesson for all knowledge, he never was able to convincingly clarify its use in QM or in other disciplines. It may lurk tacitly in the background, but it did not maintain an explicit presence in the teaching of QM. He did not satisfy his detractors, and he did not dare to cross over into a complementarity of science and faith or of complementarity within theology. Either of these may have been useful. In spite of his avowed distaste for philosophy, complementarity was claimed as a philosophical basis for all knowledge.

14. Rebellion in the Ranks

Just as Jesus found opposition within the Jewish leadership, Bohr's interpretation of QM produced negative reactions in the high ranks of important contributors. De Broglie wanted to see the wave/particle duality resolved as a wave guiding a particle. Schrödinger wanted it to be only waves, simply squeezed down to particle size. He rejected the "ridiculous" limbo state of superposition and ridiculed it with his famous dead-cat/live-cat superposition.[12] Einstein developed a twin-paradox *Gedankenexperiment* (thought experiment) to show the incompleteness of QM, the existence of reality which QM could not deal with. This work is referred to as EPR, an acronym of three authors of a famous 1935 paper, Einstein, Podolsky, and Rosen. Bohr successfully parried the blows, and stood by the strange new mechanics, bolstered by Dirac's formalism and von Neumann's attempt at mathematical foundations. With the authority invested in him as a mathematician, von Neumann had proved an impossibility theorem, the impossibility of a hidden reality undergirding the apparent probabilistic external realities. And he was wrong.

15. Where Do We Go from Here?

Von Neumann supposedly proved that a deterministic formulation giving the results of QM was impossible. Thus determinism was

12. See Schrödinger, translated reprint in Wheeler and Zurek, *Quantum Theory*, 152–67, 157.

mathematically eliminated and we would have to live with a world of chance. Using the well-known relationship of rays and wave fronts from optics and adding a nonlocal potential derived from the guiding wave ψ, Bohm was able to reformulate Schrödinger's equation as a deterministic machine. The nonlocality and the inability of the experimenter to follow the predicted deterministic paths without destroying the experiment makes it impossible to confirm. On the other hand, its identity with the mathematics of QM makes it impossible to dismiss or disconfirm if QM is correct. This means that determinism vs. the intrinsic probabilism of Nature cannot be decided. Einstein did not rejoice, calling Bohm's solution "too cheap" and Bohm himself remained committed to a probabilistic Nature, albeit one with a connected wholeness produced by the nonlocal actions. Bohm's addition of yes-no questions (of directions of spins and polarizations) made the twin paradoxes of EPR into an eventually open field for experimental studies of spin correlations.

16. Whoa! It's a Dead-end Road!

No sooner had Bohm opened the door to determinism, providing an opening back to a classical realism, when it slammed shut with the analysis of the expanded EPR experiments which Bohm suggested, making use of the two-value systems of spins and polarizations. John Bell, in full sympathy with Einstein's desire for local objective deterministic reality,[13] also proved a mathematical theorem. This theorem takes on simple enough forms that it can be understood by everyone and, even though subtle, cannot hide some concealed falsity as von Neumann's proof had. The widely known results of QM simply cannot be modeled by a local realistic model, a model where the real answers for all askable questions are specified without recourse to knowledge of the future or instantaneous communication. The idea of a limbo existence awaiting a specific question is necessary unless we admit knowledge of or communication with the future. The choice of measurement makes a difference. If that choice is free, it must be known or somehow communicated to the past to provide reality in that past time and still give the answers which QM predicts. Maybe QM is wrong? It certainly had not been tested in every possible way.

17. The Future Matters to Nature and God

Von Neumann's impossible theorem said that you could not write a deterministic history which would give the same results as QM. Bohm

13. Bernstein, *Quantum Profiles*, 84.

proved him wrong by using a reformulated but perfectly equivalent QM to give a deterministic plan which gave all of the results of QM. Bell said that a local realistic model could not give the results of QM. Experiments had not been done which would force QM to check locality. If we could really do our decisions and set-ups fast enough to require nonlocality to give QM results, maybe QM would not give the claimed results which made it appear that the decision was necessary for the results. "Which path" information destroyed interference, but we could not make choices fast enough to eliminate the possibility that Nature had found out what we were doing and acted on the information locally, i.e., with signals sent at speeds less that c. If we could delay our decision whether to look for "which path" or for "interference between two paths" until Nature had to decide which path, we might catch QM cheating. But it turned out that you can't fool Mother Nature. Nature had to use nonlocal effects, basically making decisions which used the *future* decision of the experimenter. This kind of nonlocality means that God can act on Nature, including us, from the future. A God of the past (deism) and a God of the present is also potentially the God of the future, if QM is telling us truthfully how Nature works.

18. OK, So What?

Although we are clearly caught in a directional world of progressing or regressing history, of developing or decaying life, of conserved but dissipating energy, the non-directional two-way time of classical physics may also undergird QM. Just as the mathematical waves of classical electromagnetic theory occur (mathematically) in two forms, one going forward and one going backward in time, the same concept may augment our understanding of QM and of God. Feynman and Wheeler proposed a symmetrical classical theory that included an *advanced* wave with the *retarded* wave. Cramer used this idea in a *transactional* interpretation of QM. This could take the form of a symmetry between causality (the favored child of modern science) and teleology (the abandoned Aristotelian and theological child). In theology, this would imply a God who works in the present from the future as well as from the past, a God who draws us and leads us. QM offers a different view of the world and thus a different view of the God of the world. Theology also may provide a guideline to interpretations of the physical world and its theories, if we can get our theology right.

19. Epilogue

Although I claim *your* right to write the epilogue, to make *your* decisions about what makes sense, I will try to sum up some of my conclusions. I make no claims that science will give us theological answers, nor will I try to determine scientific direction from my theology, but I will affirm that they are both of the same cloth and need to be woven together.

Chapters two and following are probably best read in sequence. Chapter one may be skipped on first reading or read separately at any time.

1

Paradoxical Issues

> A plethora of paradoxical pairs were generated by the advent of quantum physics.

THE THRUST OF THIS book is that there has been a dramatic shift in the basic ideas in the world of physics from the classical view at the end of the nineteenth century, primarily as a result of QM. We often called the development of science that took hold in the seventeenth century the rise of modern science. This continued unabated without major changes until the end of the nineteenth century. The decade around 1900 saw remarkable changes that we would now call the rise of modern physics. I think we can say without exaggeration and without undue bias that physics led the way into modern science and continued to lead the way in the twentieth century. It will sound overblown to suggest that this shift is a hinge in history, a true turning point. But it may be. My purpose is to look at these changes, which I think should be called metaphysical, something deeper and more foundational than the bread-and-butter science that most scientists do in their everyday lives. I think that the changes may affect our philosophy and theology or that our philosophy and theology may pass judgment on and contribute to the ideas and methods of science.

I am often asked and often ask myself: what are the issues that changed in the transition from classical physics to quantum physics and what issues arise in QM that might have some meaning in the philosophical or theological realm? Here is an annotated list, which is by no means exhaustive, nor are the topics clearly independent. All of these are somehow important questions. I am not trying or expecting to give any clear answers or convince you of some *correct* position to take on each of these. I only hope to convince you that they are important and they are worth discussing, exploring, and having an opinion about, even if those opinions keep changing on us all the time.

CONTINUITY/DISCONTINUITY

The first mantra of classical mechanics is: isolate an object. That is, think of the physical world as reduced to separate, individual, isolated pieces. Conceive the world as made up of *discontinuous* matter. Everything in your universe is in one and only one of your isolated objects. This may be a purely conceptual, artificial division or one may in principle get down to fundamentally indivisible pieces, a basic atomism. This discontinuity of classical mechanics is thus not a *fundamental* division, but a conventional one, a methodological one following a principle of divide and conquer (more formally called *reductionism*). The second mantra of classical mechanics is: follow your isolated objects *continuously* in space and time. This involves a fundamental belief in *continuity* of an absolute space and time and in exact laws that govern this continuous motion. There are no jumps, no surprises, no uncertainty.

QM does not deny continuity and certainty in principle. But it affirms that discontinuities, steps with nothing in between, will occur, sometimes in places or things where classical physics assumed no steps. Thus jumps, transitions over a step, will occur, sometimes as dramatic as a jump from non-existence to existence (called creation) or vice versa (called annihilation). QM often can calculate exact levels for these steps, but very often cannot calculate when or how these jumps will occur. Physics thus appears to provide a limit for its laws. Within the limits of these fundamental sizes of steps, Nature appears to have freedom with certain statistical bounds. We as knowers capable of external influences do not have sufficient or causal power to specify or yield a given effect. A determinative influence outside our ken or internal to the system must provide the necessary link *or* there is pure chance. God puts in the fix behind our backs *or* God lets a quantum system decide *or* God leaves it to chance, the roll of the dice. In some of our theological traditions, all of these conflate into one: God put in the fix even if we call it the roll of the dice or the decision of a quantum system. If that is our answer, one can hardly argue, but QM does bring the question with considerable force.

CERTAINTY/UNCERTAINTY

The certainty supposed in the classical physics view was conceived in the human mind: 1) the certainty of mathematics going back to Pythagoras, Plato, Euclid, and Descartes, 2) the presumed certainty of a Baconian inductive method, and 3) the newly invented calculus made for the seventeenth-century science of motion. A completely understandable commonsense reality and an enormously successful mathematics which described the world more accurately than we could measure it gave us an arrogant confidence. Even if we knew that 12 billion sunrises did not prove the next one and three decimal places of descriptive or predictive accuracy did not prove the fourth digit, we were sucked into belief in a perfectly mathematical physical world: all past and future were open to the Laplacean supreme intelligence who could know the present perfectly and perfectly solve the Newtonian differential equations of motion.[1] Clearly God was the model for such certain knowledge, but once we knew the mathematical techniques, the mystery of God was gone and the existence of God was really no longer necessary.

The quantum world shook us from this phony belief in certainty, determinism, and perfect knowledge. Many circumstances forced a spread of measured values on us. Identical preparations of physical systems did not allow certainty in measurements; our knowledge, even the knowledge of the supreme intelligence, was uncertain. Specifically, the two required pieces of present knowledge in a Newtonian mechanical situation were the position and velocity of each mass. Heisenberg's uncertainty principle dealt specifically with those two pieces of information; it said you cannot prepare or know *both* of these at one instant. QM gives very specific limitations, which are verifiable by experiment, for the specification of such so-called "incompatible variables." The agenda of classical mechanics and of the Laplacean supreme intelligence is dead; there is and can be no quorum of present information.

DETERMINISM/INDETERMINISM

Uncertainty is one thing, that's your and my limited knowledge, but no person of any faith at all is stopped by that; the reality and God's knowledge are something else altogether to a believer. Classical physics had led us to think that our knowledge, God's knowing, and being were all one thing, at least

1. Greenstein and Zayonc, *Quantum Challenge*, 54. This statement of Laplace represents a peak optimism in the completeness of the Newtonian formalism, published in a treatise of 1812.

in principle in the physical world. Don't limit God and the world by your own sinful finiteness, the person of faith would say. Position and velocity of a particle *exist*, are *real*, and *God knows them*, you may say (and even have Einstein on your side). Determinism is the stronger word we use for this; we can believe Nature is determined and known by God even if *we* must remain in the dark.

We have some justification for this. If we send electrons through a tiny hole, we have prepared very accurately the position of each electron which goes through. We have to admit that the velocities coming out of the hole will be scattered. *But*, we say, every one of them *has* a specific velocity and thus flies to a specific spot. Here QM gets fussy and says: it went through the hole and thus *had* a position; later it hits a screen at a certain point because it *has* a velocity, but you didn't know, you didn't measure, you can't confirm it *had* that velocity back at the hole. When and if you measure, you may give me results but you cannot tell me it *had* that value *before* you measure it.

Hard as it is to accept, that is a skepticism with some real value and justification. We can set up examples where many measurements are possible. The devastating result is that the results of each and every measurement *cannot exist,* not in a secret scroll inside the particle, not in the mind of God. If they exist *in any form*, we can show that the resulting measurement will produce statistical results inconsistent with QM, which has always proven to be right. In other words, indeterminacy or maybe indeterminability, a stronger word, must replace uncertainty. It is not just limitations of *our* knowledge. Believe it or not, you believers, God does not know all the results of all possible measurements; they are simply undetermined *unless* they are measured.

REALITY/A-REALITY/ANTI-REALITY

Classical physics had a fairly natural commonsense realism. Cartesian rationalism and mathematics made us believe in *mind;* Baconian emphasis on the laboratory and the experiment made us believe in the work of our *hands.* Classical physics made a strong link of *mind* and *hand,* producing a strong relationship between the *concept* and the *real thing*. Even in the inaccessible world of the planets, Newton's apple provided a good stand-in for planets. But quantum physics has some strong leaning toward a-reality.

Any decent physicist worth her salt will tell you that a photon does not have a polarization until you measure it. If a property or the particle itself cannot be said to exist prior to a measurement, there is a strong tendency to extend the Wittgenstein dictum against speaking about it to an agnostic position questioning its existence or to a positively negative statement that there is no reality *unless* and *until* you measure it. A strong indeterminism makes the questioning of commonsense reality a reasonable skepticism.

Once our system concedes the existence of photons, questions of reality can focus on an individual photon. Before that, on the conceptual level, we may ask if some concatenation of properties or measured realities have just been given a name, a certain kind of particle. Many times, in the world of physics, a long controversy existed about the existence of particles which were named and characterized. Photons existed for Einstein in 1905 but it took almost twenty years for the community of physics to accept it and name it. The neutrino had a similar twenty years of probation before God decided it was such a good idea that he went along with the grand invention. (Only half-joking.) In such an environment, reality lives a much more tenuous existence.

OBJECT/SUBJECT

Classical physics, with its Cartesian dualism, was pretty clear in its separation of the subject, the thinking observer, *res cogitans*, and the object, *res extensa*, the extended or spatial thing. Things *out there* do not depend on us. Science can be done in a nice antiseptic way, with a plate glass between the scientist and his prey. At least that's the theoretical point of view: mathematical and conceptual analysis does not change Nature. Baconian hands-on science takes the plate glass away and uses almost violent language in taming, subjecting, even torturing Nature to wrest her secrets from her. But there is no hint that you could get a wrong or different answer from her just because of your presence or treatment of Nature; false confessions never entered Bacon's mind. Nature has basically one answer, the truth, which cannot be trusted to be the answer provided by our totally disengaged imaginative cogitation. Neither can it be expected to be faithfully communicated through the barrier. According to Bacon, Nature will tell its truth most clearly and most faithfully under duress, perhaps duress for both object and subject.[2]

Quantum physics shatters the glass in a completely different way.[3] The object's reality literally depends on the subject's questioning. There are an infinite number of polarization questions for a given photon. Those questions cannot all have predetermined answers. When the questioner gives

2. Bacon, *Novum Organum.*

3. Peat, *From Certainty to Uncertainty*, 14. The image originates from John Wheeler.

a classic *This or that?* question with two exclusive categories, the photon *will* give an answer, an answer we have good reason to believe did not exist before we chose what to ask. The clean distinction between mind and body, a thinking thing and a non-thinking container for it, cannot easily hold. The reality, the choice, seems to require the conscious or at least sentient observer. Idealism, with reality formed by the idea, is not so far-fetched any more. Do we have a universe which is not just responsive to the mind of God, but is *in* the mind of God (and *in* the minds of his image-bearers as co-creators)?

CAUSALITY/A-CAUSALITY /ACCIDENT/RANDOMNESS

Causality in classical physics was an indestructible unbreakable chain linked from the cause to the effect. Given a set of circumstances at one time, inexorably and unfailingly one finds a particular set of circumstances at another time. We had ironclad causality. Who can argue with that? This is just what Laplace's super intelligence sees. A nice counterpoint to the theological predestination by God. God knows, God sees, it is. OK, the super intelligence knows, the super intelligence sees, it is. Since we've figured out the basic ideas, God is not so mysterious; he is just bigger in his knowledge reach, faster in his calculations, just a supercomputer self-generating universe. God might as well be dead; causality might as well be dead because there is no entry except at the beginning. The Deistic God is First and Only Cause, no more exciting than that.

In quantum physics, this kind of causality is cut off at the pass. When Heisenberg proved to his satisfaction that the two conditions needed for a classical starting point, position and velocity, could not be met, he declared causality to be dead.[4] Using the narrow view of classical causation and classical motion, he was certainly right. It is true more broadly that quantum physics does not and cannot provide that ironclad link between realities at two times.

Only one ironclad link is allowed in quantum physics and we require this to retain some semblance of our usual scientific claims. We require that measurements can be faithfully repeated. If we make a measurement, getting an answer, and then repeat the same measurement later without any disturbance of the system, we get the same answer. Fortunately, quantum nature has been that kind to us. But the more general statements of repeatability in science no longer hold. Prepare two systems identically, make identical measurements, and get identical results. *Not so!* Now, we prepare many

4. Greenstein and Zajonc, *Quantum Challenge*, 54.

systems identically by making one kind of measurement and choosing those which give a particular result. Now make a *different* measurement, taking care that the system was not otherwise disturbed. What we find are many different answers with a thoroughly predictable statistical distribution. No single result is predictable; we can find no internal or external determinant. Quantum theory is capable of calculating what answers will appear and the statistical probabilities, a nontrivial result which will have to serve as a weakened principle of causality. Our serious principle of indeterminacy means that our only recourse in interpreting our inabilities to penetrate farther is this: the system is in limbo, a condition of several potential realities until the new act of measurement actualizes it, forcing/creating a reality. The limbo condition, with a number of potential realities awaiting actualization, is perhaps *the* characteristic "reality" of the quantum world.

This is a reality with more potential, precisely because it has a weaker causality. The iron-clad chain seems to have crumbled. No script, written by God, man, or beast, or carried in the bowels of Nature, can specify all of the possible answers to all of the possible questions. God and science will have to be creative to get the answers into Nature while still hiding them from us. God seems to have done a good job of it, and science has done a good job of talking without much understanding.

CHANCE/CHOICE

There was no chance in classical physics. We can make our little games of chance: Urim and Thummin, casting lots, dealing cards, tossing coins, throwing dice; we produce chance by shuffling the cards face down or by making a complicated sensitive situation which we cannot follow in detail. But we believe it is all exactly determined by the laws of Nature and by God's command. Calvinism reigned in physics as it reigns in some corners of the religious world. I'm sure Calvin and Newton were both brilliant, serious scholars and thinkers, and their work led others to the single conclusion: God and Nature have finished their work. I at least have trouble coming to that conclusion.

Quantum physics shows a branching; at least two roads diverge in our forest. Quantum physics is saying at least a bit more; you, the participant, can set up the set of diverging roads for the systems on which you are experimenting. We certainly may decide the set of divergent roads allowed for an electron. It is asking most of us too much to call that an electron's free-will decision, but we do think of free will for ourselves. A God who predetermines what choice will be given can predetermine the choice which is made. I do not think we will know whether chance or determinism rules at the quantum level, but chance allows choice. Without chance, *a real open* potential to go down different paths, there is no choice. And choice is an important piece of my understanding of both mankind and God.

REVERSIBILITY/IRREVERSIBILITY

"Absolute, true, and mathematical time . . . flows uniformly," Newton says in his *Principia*.[5] To us, yes, it flows, and the direction it goes, we know. But when Newton's system is complete, and Laplace's super intelligence takes account of the situation, time goes backward as well as forward. Reverse the velocities and the system traverses the past exactly the same way it came. A movie of good pure classical mechanics runs perfectly well both ways; you cannot tell its direction of time. The *time of classical physics* is an abstract coordinate line which does no flowing; everything is locked into its determined time and looks more like a block of ice than a flowing stream. For the more mathematically minded among us, Newton's equations of motion depend on time twice; changing the sign twice produces no change. Time backward, time forward; no difference. You and I can easily recognize that classical physics has missed something, the time experienced by life, the time of history.

It may appear subtle but quantum physics changes that. Chance, choice, roads diverging and a decision are the stuff of life. They also seem to be necessary elements of quantum nature. The quantum world does not go backward just like it goes forward. I hate to point out where it blatantly defies a symmetry in time (because it seems so deficient, even foolish), but here goes: a measurement can be repeated *later* with an expectation of getting the same answer; a measurement cannot be repeated *earlier* with the same expectation. Our skeptical positivistic minds will not allow us to say that an earlier measurement, something we cannot actualize, would give the same answer. This becomes (for our understanding) the "measurement" problem. As far as we can see, the measurement actualizes the choice. Just as we cannot know what choice will be made, we do not have any theory of how the actualization occurs. An evolutionary process in time is described in both classical physics and quantum physics; an irreversible jump is central to quantum physics even if we do not understand it.

These decisions thus give a direction to time, a time of historical development. They serve as one-way valves for the flow of time. A chain of chances, actualized in choices, echoes the kind of time we see in a historical

5. Newton, *Principia*, 408.

relationship of God to his people. William Pollard[6] thus sees this introduction of a historical time into physics as a most significant contribution of quantum physics, just what a Christian should expect of Nature made by God.

BEING/BECOMING

Reality of classical physics centers on the *being* of a material world. Matter has mass, inertia, extension; it makes claims to its space; it endures. It is the kind of reality we can get our hands on, literally. It does not really become; it just moves around. *Becoming* requires a direction in time; otherwise becoming and unbecoming, raveling and unraveling cannot be distinguished. Quantum physics makes no claims of plans for becoming, but it at least *allows* becoming.

HISTORY AS PROGRESS/HISTORY AS DECAY

Classical mechanics seemed to be hiding something inside. The internal constituent masses of our isolated objects were ignored, lumped into total mass. The literal following of all individual masses was of course impossible. Properties that we call thermal do not seem to be matter in motion; a hot block or a hot box of gas just sits there. The field of thermodynamics grew up in the nineteenth century as an attempt to view such closed boxes without looking for details inside. Temperature could be defined and turned out to be a measure of internal energies of motion. The energies associated with mechanics, energy of motion and of pushing and pulling, had their counterparts in the unseen interiors of the black boxes of thermodynamic systems. By characterizing energy associated with temperature (microscopic motion), we could understand a conservation law, the impossibility of losing or gaining energy. What was apparently lost by one process showed up somewhere else, often hidden in the thermal energy. Conservation in the sense of saving energy sounds crazy, absolutely senseless; the universe automatically and invariably saves energy. But the bad news is that thermal processes are not always reversible. If we use our macroscopic motion or work to heat up something, we often would like to take that energy back in its macroscopic useful form, say to move a train. The second law of thermodynamics gives us the bad news that Nature exacts a tax on that process. Microscopic processes may store all that energy, but it will not give it all back to macroscopic

6. Pollard, *Chance and Providence*, 71, 97, 104, 113.

energy. There is now a direction of time in physics and it is going downhill; energy does not go away but it eventually all gets downgraded into useless thermal energy. This we call the heat death of the universe. We have billions of years left, but we are doomed.

This principle began to be understood and clearly enunciated in the 1850s, almost exactly when biology began to clearly enunciate its principle of salvation, Darwin's *Origin of Species* of 1858. Life and evolution (which is life on a larger scale) produce complexity or higher forms of energy. Life is precisely what is capable of using the material and energy environment to grow, reproduce, to make its world more ordered, raised to a higher quality. This seems to be a contradiction and an ultimate face-off between biology and physics. It is and it isn't. The second law of thermodynamics applies to closed systems, into which no outside material or energy can come; we believe that it thus applies to the universe as a whole by definition. Life (and evolution) on the other hand requires an open system; it has to eat to live. We, the living, make our world "better" at the cost of the environment, the rest of the universe.

So science has *two* time-directional laws, based on life and death, clearly at cross purposes. We know both in our personal experience and I think we can see both in a longer history of mankind. They are both in our worldviews, and they are not easy to understand in our theology. Think of the fall, heaven, and hell and see if they fit one of the scientific time-directional laws.

RELATIVISM/ABSOLUTISM, NARRATIVE/ METANARRATIVE

Classical mechanics promised a complete and absolute metanarrative, the whole story open before the eyes of Laplace's super intelligence, who just implements classical mechanics. To a large extent, we are still swallowing that story. On the other hand, we hear and fear the competing story that *all is relative*, that all narratives have their truth, that no narrative can claim *the Truth*. Both views gain some credence from *absolutely false* understandings of physics. The classical framework was never as solid a framework as we imagined and it has clearly come crashing down. Relativity and quantum physics are the legitimate key revolutionary developments which capsize or supersede or negate classical physics, and the key words, *relative* and *uncertain*, are enough to shake our faith in an absolute and certain foundation and send us into an orgy of pluralism, diversity, tolerance, anti-foundationalism and postmodernism. Without denying the value of such components of our new (oftentimes unconscious) metanarrative, we need to note that they are not justified by the sciences so often used (even tacitly) to bolster them.

The uncertainties of our limited knowledge are pounded into us in the quantum world. There is no question of a loss of certainty for us. Try as we might, we still cannot rule out certainty underneath it all. Relativism is a complete misunderstanding and misnaming of what goes on in the theory of relativity. The absolute *space* and *time* of Newton's world is shattered, and Einstein exorcized its material impostor, the ether. The *relativity* of meter-sticks, clocks, masses and simultaneity blows our minds, but it is based on the *absolutes* of the speed of light and the laws of physics. Although it was not in the title nor was it prominent, Einstein did use the word *relativity* in his original paper of June 1905. To annihilate the troublesome concept of absolute motionlessness, he proposed that all mechanical, electrodynamic and optical laws must take a form which is independent of *relative* motion. Stating it in the negative, he proposes a "principle of relativity" (*Relativitätsprinzip*): the laws of physics *do not depend on relative* (uniform translational) *motion*. Stated positively, the laws are *absolute*. As Einstein would have preferred, the principle is an *Absolutprinzip.*

The unfortunately named theory of relativity thus dramatically relativizes what we held dear as absolutes, space and time, in favor of other absolutes, the speed of light and the forms of the physical laws. Clocks run at *different* speeds, and meter-sticks have *different* lengths, precisely because the speed of light and the laws of physics are not *different* for observers moving uniformly with respect to each other.

By its indeterminism, its probabilism, its limbo states, QM only adds to the loss of certainty. The existence of uncertainty is not relative; all observers experience it. But again, deep down in the heart of the quantum world, a new absolute appears. A new constant, h, called Planck's constant, rules the *limits of certainty* on the small scale, just as c, the speed of light, rules as an *absolute speed limit* in the world of high speeds. As crazy as the limitations of our classical physics appear to us, a loss of absolutes is not a concern.

For Christians who are told that metanarratives carrying any absolute claim are *verboten* in our postmodern pluralistic tolerant world, the absolute limits held by our "postmodern science" support the possibility of the Christians' claim to an ultimate solid unshakable foundation. Science can only find those *physical* absolutes, which can only point to possible absolutes in the realm of life and spirit.

SINGLE DESCRIPTION/COMPLEMENTARY DESCRIPTIONS

Classical mechanics had a pretty solidly unitary description: we had masses, which were the stuff of the universe; we had forces which acted between them; the forces on each mass provided thoroughly determined changes in motion. Masses we could touch and heft; motion we could measure by our sight. Everything was sense dependent and had simple correlates between reality and sense data. Well, almost. Forces of this simple *feeling* variety, push and pull, require touch, but one of Newton's great triumphs was to realize that every mass in the universe pulls on every other mass in the universe. It is precisely when we are *not* touching the earth that the earth's pull on us is of great concern; we will fall. Action at a distance, a non-contact force, took forces out of the commonsense reality. Newton neatly swept it under the rug, by writing a nice mathematical formula for it. He described its *quantity* well but refused to hypothesize how it occurred, i.e., to give a commonsense/mechanical model. When electric and magnetic forces acted similarly at a distance in the next few centuries, Faraday, a *much* lesser mathematical mind who got his scientific training by way of an apprenticeship as a bookbinder, needed a visual model of what could be happening. He envisioned "lines of force" to represent what we now call fields, something filling space to reach across from one charge to another (electric fields), from one moving charge to another moving charge (magnetic fields), or from one mass to another mass (gravitational fields). Maxwell, using the visualized lines of force and a great mathematical mind, built the concepts of fields into a magnificent structure, a go-between allowing masses and charges to act at a distance. Masses and charges produce fields which act respectively on masses and charges. This made some sense but fields needed *something*. We gave them the ether, a space-filling, very stiff but very thin medium, *ersatz* space, a minimal mechanical scaffolding for the mechanically minded physicist who could not give up the matter-minded view given by our senses. When Einstein kicked the ether scaffold down, the fields, both the standing framework of statics and the wiggles of the dynamics, were still there. Thus was invented, as a capstone of the classical mechanics, a dual picture: matter *and* force-carrying fields, matter existing in clumps, discontinuous pieces, *and* fields filling all of space, a continuous connector. Fields had that wind-like quality and we can reasonably be reminded of the Spirit hovering over the waters in Genesis 1 and Jesus's description to Nicodemus in John 3 of those born of the Spirit.

This final picture of classical physics gives a mixed picture of unity and duality. The two pieces, matter and fields, are so different and yet the field is the glue that binds all the matter together, an imposed unity. Moreover Maxwell's theory bound electric and magnetic fields into one, an electromagnetic field. It also showed the unity of electromagnetic fields and optics; optics consisted of propagating waves of electromagnetic fields. Einstein's vision for a unified field theory was supposed to also tie in the gravitational field, a vision that served as the elusive pot of gold at the end of the rainbow for the ageing genius.

It was the genius of Einstein to jump back to the *discontinuity* for optics, in spite of his enormous respect for Maxwell and his electromagnetic wave theory. Light for Maxwell was an electromagnetic wave. Fine, says Einstein, but it comes in pieces rather than in a continuous wave. It became the genius or the Achilles heel of QM to recognize that our discontinuous *matter* world also partook of that same duality/unity; all things carried both natures in their unity. "If it were only a Trinity," immediately responds our orthodox belief. It of course won't do to call such a unity in duality absurd in physics and then use it as the analogy for God-man or for the Trinity, as some sincerely believing legitimate physicists do.[7] QM may be limping, but it is walking on the two legs of complementarity, Bohr's main principle of quantum interpretation. This principle may in fact pervade all knowledge as Bohr preached. Wave/particle duality *and* a reality of wave/particle unity are both pervasive concepts.

LOCALITY/NONLOCALITY

The questions of discontinuity and continuity in a particle/field duality sound like a concern of locality: does a particle stay where it belongs and affect only those it touches? With a concept of field effects capable of propagation, locality takes on a much more detailed and technical meaning. If a star is born, it may well remain confined as a ball of burning gases in a small region of space. However, if it is accompanied by fields which propagate from its birth-time in all directions at the speed limit of the universe, it has an ever-increasing locality, a literal sphere of influence. One year after its birth, it can "touch," i.e., have gravitational or electromagnetic effects up to one light-year away, almost 10^{16} m away. Fields can solve your conceptual problems of how to reach out and touch someone at a distance remarkably well; just send your effects out by fields at the speed of light. This would seem to solve all action-at-a-distance problems for all practical purposes; it takes about 8 minutes for the sun's light to reach us, but the sun was there 8 minutes ago; we are thus local to the sun. Insistence that action at a distance,

7. Hodgson, *Theology and Modern Physics*, 125, 171.

a cause-effect relationship, must be *local* means that the effect cannot be outside that sphere radiating from the cause at the speed of light. This is a reasonable proposal which you would expect Einstein, the original speed-limit cop of the universe, to insist upon.

The Copenhagen interpretation of QM, Copenhagen being the seat of Bohr's "papacy," basically requires Nature to break Einstein's laws, i.e., to act *non*locally. Einstein realized that the measurement problem was sufficient evidence for such a crime and set out to arrest QM. Actually, of course, he did not believe such a crime was possible, so he claimed that QM as it stood in the Copenhagen interpretation must be wrong, or as he more gently put it, incomplete.

To show this, he introduced a simple thought experiment involving two particles separating from a common origin. From the common origin, they have properties which are "entangled"; knowledge of one implies knowledge of the other. (For example, they are twins and have the exact same properties, or they are anti-twins and have exact opposite properties.) Since many properties could be measured, and QM says a system is in limbo and does not possess the property until measured, we have a problem. They are separated but knowledge of one implies *instantaneous* knowledge of the other. A local observer-created reality on one side becomes a nonlocal observer-created reality on the other side. The limbo is broken instantaneously at a distance, what Einstein called spooky action at a distance. To soften the blow of such an inconceivable idea, some have called it passion at a distance.[8] Einstein's claim was that the supposed limbo state of Copenhagen interpretation was in fact a reality of what was to be measured in the future, a reality that QM did not know.

Pause . . . . At least for a WWII and an atomic bomb and many years besides. Who was right in this gigantic war of the reality of the underworld of the atomic level? I guess I've already told you. The reality of what QM calls limbo produces logical impossibilities. The only way out is to admit nonlocal influences. Causality, presumably dead at the hands of probabilism and uncertainty, comes roaring back with a vengeance; *impossible* causal effects must be occurring. The world is acting nonlocally, whether you call it spooky or not.

REDUCTION/WHOLENESS

A commonsense reality of classical objects, objects that were separable, that in fact cannot occupy the same place, is an appealing reductionist world.

8. Shimony, *Search*, 133.

The pieces can be fit together or taken apart. Following all the parts and thus seeing them in their variety of relationships seems like a pretty complete picture. Quantum physics has numerous ways of breaking that simple picture and exhibiting wholeness.

First, the particles are *not* isolated points. This is the lesson shown by their wave nature; waves go everywhere into all the nooks and crannies and two waves can co-exist in the same place. This is also the lesson of a limbo existence; a quantum does not go through *this* slit *or that* slit but through *both* in its limbo potential existence. You may force a particle to answer the question of which slit it goes through, but if you don't ask the question, it does not volunteer any information. It acts as if it used both paths, or an infinite number of paths, if you let it.

Second, particles that have interacted with others in the past have an entangled future, a *correlated limbo* which requires faster-than-light signals, spookily connecting particles to their old friends and enemies. If one can call this a particle, it is an impressive particle; it can go anywhere and still know what is happening to its past associates.

Third, a particle is totally indistinguishable from others of its ilk, and may be switching identities with them at all times. We learn this last manifestation of wholeness from what we call quantum statistics. Classically we learn about statistics by imagining a number of similar particles or systems with which we can enumerate all possible combinations of properties, as though each system has its own identity. For example, the *first* die of a pair of dice can give each of six answers, as can the *second*. There are thirty-six combinations of properties. This gives us a classical reductionist calculation of probabilities. What we find in quantum physics is that *no* such *distinguishable* fundamental particles exist and there are two types of *indistinguishable* particles which we call fermions and bosons. The difference in the two particles lies in what happens when two particles trade places. Fermions trade places in such a way that it is impossible for two of them to be in the same state. This gives them properties we associate with real particles in our macroscopic world which cannot occupy the same space. Thus these particles fill spaces in the sense that there is no room for other particles there. We can call them anti-social; they each get their own space. These are particles like electrons, protons, neutrons, etc., which make up what we know as the *material* world. Bosons are somewhat opposite. Their method of trading places encourages company; they effectively want more bosons to occupy the same state. Both bosons and fermions effectively know every one of their counterparts in the whole universe, a dramatic show of unity and wholeness.

The so-called measurement problem, the insistence of a system to wait for a measurement operation before choosing the reality, is another form of the wholeness of the universe, an inseparability of what we want to call the object and the subject. The plate glass of subject/object separation is shattered forever. The veil in the temple was rent in two. To declare independence of particles from each other is to do violence to the picture of Nature; to declare our independence from God is the original sin.

CAUSALITY/PURPOSE

The causality of classical physics was an imprisoned causality, a determinism locking us and God into a lifeless block universe where things did not *happen*. Things just *were*. To talk of God, you had to get outside the block; he *made* the block. You may call one end of the block the beginning and the other the end, but it didn't *go* from beginning to end. It was thoroughly known by Laplace's imagined super intelligence, who knew nothing but classical mechanics. Eternity as a block of time, albeit infinite, rigidly connected from beginning to end is a weak caricature, even if the block can be curved and shaken like Jell-O by Einstein's general relativity. The whole system becomes a caricature of God, Nature, and causality. Purpose was explicitly driven out, and such a Nature does not have a purpose; it does not fly.

Quantum physics chipped away at that overly simplistic picture, but it has not yet put a coherent picture in its place. A renewal of purpose, a return to teleology, could go a long way toward coherence and completion. Causality of classical physics with its deistic view of God the Creator as merely an initiator of the universe has always seemed deficient to me. Physics itself has often played with concepts which would provide a correction. The Maxwellian wave equation for electromagnetic waves is, along with Newtonian mechanics, also time reversible; waves go backward in time as well as forward. The God of our past is also the God of our future. We are not just driven by our past; we are drawn to our future. How our present is determined, even *if our present is determined*, may be and remain an unanswerable question. The theological symmetry of a driving and a drawing, of causality and purpose, of origin and *telos*, of First Cause and Final Cause, of beginnings in Eden and ending with the Marriage Banquet, of the sheep dogs behind and the Shepherd ahead, all have drawn me to a physics which recognizes a causal future. In physics, it naturally takes lifeless-sounding forms: backward causation, retro-causality, *post*-paration (instead of *pre*-paration), advanced waves, or time-symmetric quantum physics.

Two centuries before the harmless backward waves considered to be unphysical appeared in Maxwell's electromagnetic theory (nineteenth century), three different theories of light gave possible illumination to the modern dilemmas (trilemmas? now that we have the trinitarian three?). Any textbook on light should give you a good introduction to these three ideas: 1) wave fronts, 2) rays, and 3) the path of least time. The first clearly looks at waves, the second gives a particle picture, and the third takes a holistic view which does not care whether you think of particles or waves. Huygens was responsible for the first picture; simply produce a wave front from the previous wave front by letting each point on the wave front send out a wavelet. The ray picture just imagines a particle-like ray, a corpuscle in Newton's language, which goes along a path perpendicular to each wave front. The third method (due to Fermat) basically says Nature tries all the paths from a beginning to an end point and takes the shortest (in time) path. Waves and particles obviously can be complementary pictures of the same phenomenon.

The principle of least time has excited some physicists in the successive centuries because it clearly connotes purpose, an ability to seek out the best of all possible worlds. The skeptic will tell you that this supposed purpose is just a mathematically transformed system and does not do anything like our anthropomorphic description of *trying all the paths and choosing*.[9] QM gives us an impetus to rethink this for two reasons: 1) the limbo concept of somehow being in all states or paths at once without determination and 2) the spookiness of nonlocal connections, instantaneous communication. Waves (or signals) backward in time allow these to make sense. If I send messages forward in time and my friend sends a response backward in time, they, the message and response, are for each of us instantaneous while an outside observer sees the appropriate delay we associate with a propagating signal. An entity going forward in time, then backward in time takes no time for the round trip and thus can take all the paths simultaneously. The measurement problem and the question of reality are all solved by this simultaneous causality/purpose. A measurement forces the choice but the choice is not then at the last instant because it is communicated backward in time.

Skepticism is understandable and fine. The organismic connection of all parts both in time and space, a meaningful purpose and becoming are appealing images compared to the deterministic reductionist machine. Life, the church, and world are organized as interactive acting bodies, not as cogs in a machine. Independence and freedom, the mainstays of American

9. Margenau, *Nature of Physical Reality*, 422.

thought, can be ugly isolationism and competition. As W. H. Auden put it: "The slogan of hell: eat *or* be eaten."[10] I hope that you can guess and understand what his "slogan of heaven" is: eat *and* be eaten. The Declaration of Independence for Americans needs to be replaced by the Declaration of Dependence.

10. Auden, *Certain World,* 229.

2

Seeing beyond Seeing

> God has broken through in incarnation, and radioactivity broke open an otherwise invisible world deep in Nature.

On November 8, 1895, a full week too late for Halloween, Wilhelm Röntgen saw a ghostly image (pictured below) of his wife's hand showing the bones like a skeleton with her wedding ring showing clearly on the appropriate finger. We would not be surprised or shocked by this because we have seen many X-ray pictures. But this was the first time and it sent a shock around the world, especially to the physics community, beginning an absolutely new era.

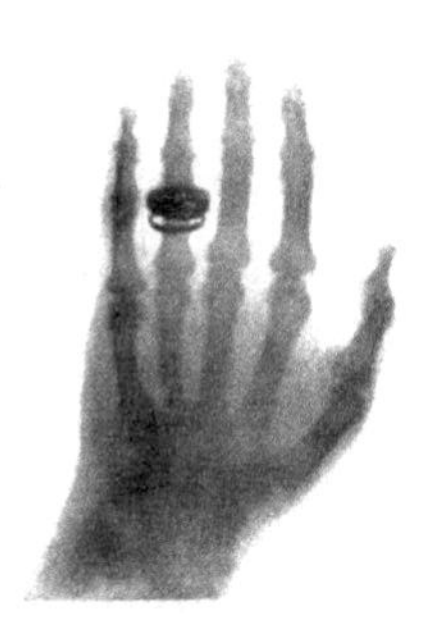

Seeing beyond seeing, that's my definition of both science and religion. We are always seeing things, events, connections, or images that beg for us to see behind, under, around, or through. Christian theologians did not come up with an idea of a Trinity because three was some magic number that should represent completeness, or community, or perfection. We experienced a creation that we felt must have been made by a creator. The Israelites experienced an historical interaction with a covenant God, a law-giving God, a God-king, and a jealous God who insisted that only one God should be worshipped and obeyed. Then the Jews (and later the gentiles as well) experienced a miracle-working, impossibly demanding Jewish teacher who died at the hands of the ruling empire, rather than physically overthrowing it, was resurrected, and claimed unity with God. When he left, he promised that he would send the Holy Spirit who would continue his work, and a church was born which experienced the power of that being. Our best God-theoreticians (whom we call theologians) have tried to piece together a model, a picture of who or what God is, this source of all these experiences; this subject becomes our object(s). That such an immense Ruler of the Universe, as he appears to us, should continue to *choose* to come to us is what we call *grace*. Just as we rely on our sensual experiences to give us categories to name our latest fundamental particles by color, red quarks, etc., even though a quark cannot even be seen or even separated from other quarks, we might think of colors of grace. Green grace is the grace of creation (the color of grass); red grace is the grace of incarnation and redemption (the color of blood); blue grace is the grace of sanctification (we probably don't even know why blueness is associated with suffering and sorrow).[1] God produced us and all we have; God came to be with us; God stayed with us.

Pardon what seems like an excursion when I'm telling a story about Wilhelm Röntgen, the first Nobelist in physics. It is in fact the other way round. Our story is the story of grace, and our tales of science are just little excursions into what we have experienced in our physical realm which may help us understand what we've experienced or can experience with God. I have been fortunate to experience God in both science and my Christian faith. When I try to see them whole, as one unified experience, I find that my experience and the ideas it generates seem foreign to those who experience science or faith as a somewhat separate isolated world. Thus I am trying to reflect out loud on my experience. Back to Röntgen.

Röntgen was working on "cathode rays" which were shortly to be called electrons. They apparently streamed from one electrode to the other in the electrical discharge of a sealed glass tube filled with low pressure gas. The beam of electrons could produce visible evidence on fluorescent, phosphorescent, or luminescent materials. We are all familiar with similar things in neon lights and fluorescent tube lighting. All early TVs and computer monitors were fancy tubes of this type (called cathode ray tubes or CRTs) by which the electron beam is manipulated in intensity and position to write words or draw pictures on a fluorescent screen. The *Originalröntgenstrahlungsapparat* (original Röntgen-ray apparatus, in English) was a simple tube with a slanted metal electrode for the electrons to strike.

1. Simmons, *Entangled Trinity*, 178.

To operate, simply connect an electrical voltage of thousands of volts. Volts are familiar to us, with batteries commonly providing voltages of 1.5 V up to 12 V. This is because an electron process in an atom requires or can provide energies of this size, which we call eV, electron-volt. Volts are for charges what elevation is for masses. Climbing a high hill requires an outside input of energy for a mass; raising a charged particle to a high voltage requires a source to give it the energy. Sending an electron through a high voltage is like dropping it off a high cliff. Ten volts is enough to knock an electron off a typical atom; a voltage of 10,000 V gives an electron enough energy to knock electrons off of 1,000 different atoms as it tumbles down the cliff. A voltage of 10,000 V was typical of a CRT in a color TV. The atomic processes often produce light we can see. But what happens behind all this? What are we *not* seeing?

This question led Röntgen to cover his tube in a sealed black-carbon container, work in the dark, and use a plate with a fluorescent coating to try to see any invisible energy. He could also use photographic plates to take pictures but that required processing. The fluorescent screen can give a picture immediately (as a TV screen does). He saw activation (glowing light) with his plate as far away as two meters from the covered tube. That meant that invisible energy was going through the container and through two meters of air, and still produced light over the large area of the sphere of radius two meters. Then the eerie skeleton of his wife's hand appeared when her hand was placed between the CRT and the detecting plate. This was the first *Röntgenogram*, what we and the modest Röntgen called an X-ray image.

Röntgen spent several weeks without leaving his lab, wrote a ten-page paper (without pictures) over Christmas, and had ninety envelopes addressed to scientists all over Europe ready for mailing on January 1, 1896, nine of which had photographs for his close friends or distinguished scientists. From a former student, to a friend, to the friend's father, the editor of the leading Viennese newspaper, and by January 6, the German press had spread the "sensational discovery" to the world. Being able to see through clothing was sensational in the Victorian era, at least in England where a London company made a fortune on "X-ray proof clothing . . . for the sensitive woman." The sensation spooked its discoverer; he gave only one public address, not even giving a Nobel lecture. He returned to less sensational areas of research. Medical applications were immediate with usage occurring within weeks even in the backward former British colony of America.

Henri Becquerel was a third generation physicist who had inherited a collection of uranium salts from his father. These were phosphorescent, emitting a slowly dying glow upon being activated by bright sunlight. Becquerel mistakenly thought the mysterious X-rays came from such activated glowing materials and tested his salts within weeks of Röntgen's discovery by exposing them to bright sunlight and laying them on well-covered photographic plates. Sure enough, radiation penetrated through the covering into the plates and left images of the crystals, even penetrating a copper plate in the shape of a cross to leave a cross image on the photograph. Then he had *his* accidental epiphany which exploded the newly exploded science. The sun didn't shine for several days and he could not activate the phosphorescence. But the penetrating rays persisted; they were coming from the *unactivated* uranium salts. They had their own internal, apparently eternal, activation. He had inherited them from the prior generation, but his father had obtained them from the Earth, which was maybe 100 million years old (the physicists' outside estimate at that time). The complete story of *Becquerel rays* (or *les rayons uranique*, as he called them) is way too long and involved for our present purposes. Becquerel shared the 1903 Nobel Prize in physics with Pierre and Marie Curie; thus four of the first six Nobel laureates in physics had worked on these invisible rays (the prizes in 1901 and 1903). The work in radioactivity continued in the Curie family (the daughter and son-in-law winning the Nobel Prize in chemistry in 1934) and in science. You know the result, *the bomb* in 1945 almost exactly fifty years after the radioactive bomb had changed physics completely. God, with his natural world, had whacked us up the side of the head with a two-by-four when we had the lights turned out, shut our eyes, and turned our back. The energy in these new rays was about 1,000 times that of X-rays, making them about one million times the chemical energies involved in atomic processes. It is little wonder that individual events (one ray from one atomic event) could produce effects that we could see or hear, a light flash, a click of a Geiger counter, or even trigger a bomb.

Why do I tell you this story of radioactivity as the earthquake of science? It introduced us to the quantum and to the major feature of quantum physics, chance.

The word *quantum* basically means *how much*, a long-term feature of science, the quantitative measure. By the end of the nineteenth century, leading scientists could boast that measurement, the ability to attach a number to something, was what made it science. As Rutherford scornfully observed, otherwise it is just stamp-collecting. We may agree with them that quantitative measure, the mathematization of our knowledge, is a powerful tool. But we should also remind ourselves that the things that are really important in life, such as love, grace, etc., are *not* quantifiable. Quantum physics is not just quantification, but deals with discontinuous, discrete quantities, situations that do not allow continuity but require or provide a jump. Before the advent of radioactivity and its eventual boost to the concept of atomism, we did not see individual atoms or atomic events. We saw averages of many events that were so small we could not see them individually. Calculus is based on limits and assumes that any two numbers, no matter how close, have a number between them, and classical physics assumed that all of Nature behaves the same way. Two points in space always have a point between, etc. But radioactivity had large events that were large enough that we could know there was not a comparable event in between. In the early days of radioactive research, graduate students literally sat in a dark room and *counted* flashes of light. Thus it appeared that some quantum measurements could only be in *integers*. When we measure a length, we think we could always use a smaller scale to get more accuracy: from meters to millimeters to micrometers to nanometers to . . . with no limit *in principle*. But quantum effects are like ladders with no place in between the rungs. You do not climb up 2.45 rungs on a ladder, just as you do not have 1.8 children. That jump from one to the next, with nothing in between, is the key feature of quantum physics. The mathematics that we associate properly with classical physics, the calculus, which is usually only approached at the college level and is a barrier to most people, is reduced in the quantum world to *counting*, which is accessible to three-year-olds: 1, 2, 3, ... And radioactivity was our introduction to that new world.

The radioactive rays did not at first seem so powerful. Pierre Curie actually described them as "weak rays" in his Nobel lecture (1903), because there were so few of them.[2] Uranium salts had been doing this for millions of years, so the amount we would see in a few weeks would not seem like much. But the penetrating power and the lesions caused by carrying radium in your pocket (he also tells this in his Nobel lecture) were more than hints of the great power buried deep inside the atom. Radioactivity from uranium, thorium, radium, and actinium seemed to last forever with no measurable diminution in a few years. But soon radioactive materials, especially radioactivity *induced* by the primary sources, were found which were clearly unstable and died (or decayed, as we call it) in reasonably short times. You can even do three-hour laboratory experiments capable of seeing the decaying radioactive rates, literally counting the numbers of events in the thousands. When you do so, you find an important fact: spontaneity or chance or atomic free will.

In Curie's Nobel lecture, the events were described as *spontaneous*. That means you cannot predict the events; they seem to occur by chance, choice or free will, i.e., without cause. In one sweep, causality, predictability and reproducibility are out the window. Science, by our junior-high

2. *Nobelstiftelsen* (Nobel Foundation), *1901–1921*.

characterizations of it, is impossible, but you can see that has not slowed science or diminished our culture's high regard for it. Chance, if you ask any modern physicist, is built deeply and inextricably into Nature. If you are disinclined to believe that, I do not hope to dissuade you, but I will tell you some of why the radioactive phenomena make us think so. And, let me warn you, this was just the beginning, the first salvo. Physicists were already quite facile with statistics and probability, because we played games of chance and because the huge amount of data associated with atomic phenomena could only be handled statistically. Even with present-day phenomenal computer speeds and computer data storage, computers cannot handle even small problems within the lifetime of the universe if they are required to keep track of all atoms.

Consider a collection of radioactive atoms. Try to measure, from your arbitrary starting point in time, the life expectancy or *lifetime* of the atoms, the time required to produce a radioactive ray and thus die. We call it a decay, a transformation which changes the atom and gives off a ray in the process. Measuring the time for many atoms gives three remarkable results:

1. They appear spontaneous, i.e., there is no pattern or causation which can be found;
2. The average lifetime obtained is independent of when you started. The sample may have been there for all earth's history or may have been prepared yesterday. The past history does not affect the life expectancy.
3. The standard deviation of the lifetime is the same as the average. You often hear measurements quoted as an average, plus or minus the standard deviation. For example, I may reasonably quote my weight as 180 ± 2 pounds. A radioactive lifetime measurement might give, for example, an average of 42 minutes, but the uncertainty (as measured by standard deviation) will also be 42 minutes. Thus we would say the average lifetime is 42 ± 42 minutes. If I told the drivers' license office that my weight is 180 ± 180 pounds, they would throw me out. Thus physicists will measure and quote the *half-life* of a radioactive sample without bothering to note the embarrassing uncertainty in the lifetime of an individual atom. (Half-life is the time required for half of the atoms to decay.)

Stated in this way, we should hope that more than a century of work would allow us to do better, to get a smaller standard deviation and/or find some causes which might be named, measured, or controlled. We have not. When you realize that the energies involved are 1,000,000 times that of ordinary chemical or light processes, you can see that changing properties such as pressure or temperature, even in extreme levels like the sun's surface temperature, will not have a chance to make a measurable difference. Another way to say it is that we would have to penetrate into distances down to one-millionth of the size of the atom. We can of course produce these kinds of energy with the huge present-day accelerators, but when we do, we just blow up the whole inner world of the atom. We do not better understand the natural unstable process of radioactive decay, at least not in finding or controlling causes or predicting an individual atom's time of decay.

There is a reason that neither science nor God need blanch at the prospects of such lack of knowledge and control. With very large numbers, which is the level we deal with as macroscopic beings, the statistics work out almost perfectly. If we have 2,000 radioactive atoms, about 1,000 atoms will decay in a half-life; more carefully stated, it is 1000 ± 30 atoms, i.e., we know the number within 3 percent. For the enormous numbers involved in even a handful of a pure substance, the uncertainty will become unmeasurably small. When you deal with a single atom, our uncertainties and lack of causality are still there in all their glory.

This is not a negligible problem because a single radioactive decay can be used to trigger a nuclear bomb. Such an inscrutable process with such ill-defined numbers seems impossible to model, to begin to understand. However, if one allows chance, it turns out to be a very natural process. We can start with a very simple model, a coin flip. Suppose a radioactive atom has a 50–50 chance of decay in one time unit; we can simulate decay as getting heads in one coin flip (our unit of time). Try it, as I did. Trying 16 atoms/coins, seven "lived" only one time unit (they came up heads on the first toss), 4 lived 2 time units, 3 lived 3, 1 lived 5, and 1 lived 6. That gives an average lifetime of 2.19 time units and a standard deviation of 1.47. The following histogram shows the decay of the entire sample.

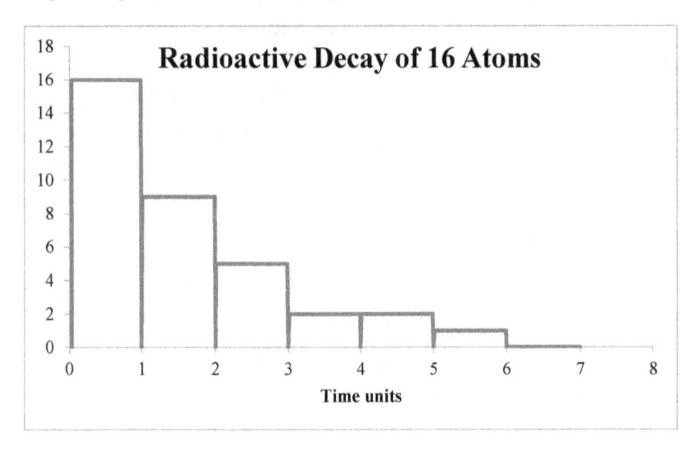

You can see that this does not exactly meet our criterion (3) that the standard deviation equals the average. If you do this for *many* more coins, you will still not produce that. (You will get an average lifetime of 2 and a standard deviation of 1.414 = square root of 2. It's easier to work theoretically than flipping thousands of coins.) But if you shorten the time for one trial and make each very short time have only a very small probability of decay, you will produce the curve that any large sample will give, an exponential decay (shown below).

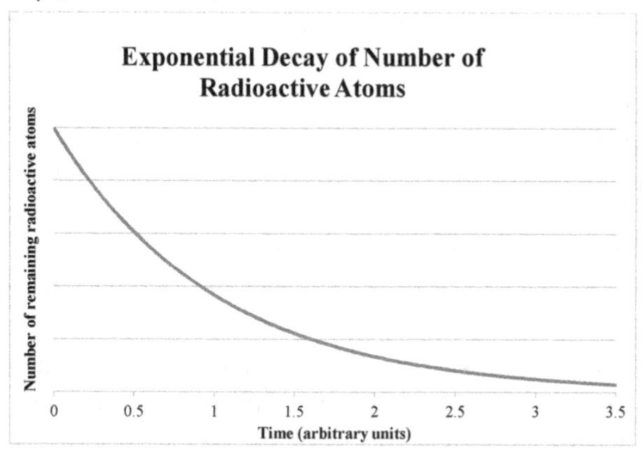

Note: we are using *decay* in two common ways, but they may be confusing. Each atom or nucleus decays to produce a radioactive ray, but the size of the sample decays by the loss of atoms capable of decaying. The sample size decay is what we are showing here. Note that the shape of the curve and the half-life are independent of where we start. This is a well-known common mathematical form in physics. For example, your piping hot coffee cools to room temperature with a graph like this.

This means that a reasonable model for radioactivity is that each atom is playing the lottery and becomes ineligible once it wins. A few decades after the discovery of radioactivity, George Gamow proposed the crude idea (and quantum mechanical calculation) of each ray bouncing back and forth in a nuclear prison with a very small chance of escape each time it hits the wall. Such a fair random lottery mimics our experimental results.

For those of you who chafe at the idea of chance in God's world, let me take your side for a moment. God could know and cause every seemingly random process. The casino operators could rig and/or know every roll of the dice, spin of the wheel, deal of the card, etc. If it is random *to us*, it means *we* cannot know the cause. Einstein famously took your side and could not believe that God played dice: *Der Herrgott würfelt nicht* (God does not play dice). He was not concerned about the supposed immoral or careless action of a personal God, but he was concerned about the production of orderly laws of physics out of the apparent chaos of random processes. Stephen Hawking said that not only does God play dice, he hides it from us. If he plays dice, he is hiding the process from himself and does not need to hide it from us. He may cause and know every apparent throw but make it look random. He may throw dice, know the results and edit to his heart's content before releasing the results, again maintaining the appearance of randomness. Or he may actually throw the dice and accept the result, without foreknowledge but with full assurance that his overall design and overall control are sufficient. It is my belief that casino operators really don't know the results and don't rig them. They easily get their results (making money from you or me) without that kind of cheating. If the world ultimately continues to show an appearance of crap-shooting, our theology will need to answer it. Is it an accident that *dice* appears in theo*dice*an? Both God and Nature, if you can or wish to separate them, have managed to do quite well with a lot of incredible order in the midst of a world of chaos, apparently based in the very small on lots of dice-throwing.

God is love. Love is vulnerable. Vulnerability means taking a chance. This summarizes the whole history of our interaction with God. He took a chance in making the universe and mankind. He took a chance in coming in the form of man to redeem us in the death, burial, and resurrection of Jesus Christ. He continually takes a chance in being with us through the Holy Spirit. It is dirty nasty work in what is clearly a lot of nastiness and rejection. Our claims to civilization and culture which presumes that we could do better than God and/or the uncivilized past or isolated savages should be pretty well disconfirmed by twentieth-century history. Those who continually state that "we now know better" are also reduced to saying that "that kind of action does not belong to today"; in other words, our lives are full of confessions that we have not achieved what we dimly see and envision as the right way. What we see is a world consistent with a God who gave rules which were broken, but by his mercy and grace we have still survived.

Why is this world so bad, if God knew what he was doing and was perfectly capable of making it according to his choice? Who am I to try to answer such questions? But here is my attempt. God had choice and he wanted mankind and Nature to have choice as well, to be like him in that regard, so that we matched with him, could relate to him and him to us. This had to be *real* choice, not a rigged system which looked like choice. When there is real choice, there must be the opportunity to take different paths, bad ones as well as good ones. Murphy's law says that if it is possible to take a bad path, someone will do it. And we did and we do. I don't know what paths are good or bad for our electrons, X-rays, etc., but it appears that God gave them a choice. Since we cannot meaningfully consider them as choice-makers, we can only describe it as chance. Nature at its core is certainly inscrutable, just as God is. That's why science and religion, both trying to peer into the unknown and unseen, can seem so much alike, both so proud of what they have and know but also so in awe of the mystery around every corner. Richard Dawkins is a religious person, a strong believer; he just believes in a god that he will not identify with the gods of the great religions.

All of our preconceptions which led us to science and our post-conceptions which come from our science must relate positively or negatively to those of our religion. This book is my attempt to see how one major development, the rise of quantum physics, may help us to see the grandeur and majesty of what we can see but also the abyss and mystery of what we cannot see or know. I hope that knowing a little more of how science has developed can be an avenue of grace for you.

3

God, Classical Physics, and Modern Science

> Kepler and Newton (among many others) built a science in a world of Western Christianity, with roots in a Christian worldview; the worldview of classical physics and the worldview of the Western church became often indistinguishable.

PHYSICS IS A WORLD of arrogance. We in physics tend to think that we are *the* fundamental science to which all else can be reduced. Religion is just social science, which is just psychology, which is just biology, which is just chemistry, which is just physics. QED. Never mind that there are some missing links along the way. Physicists can often think this way, but society outside of physics has just as often placed them on that high pedestal. The result is that our culture has assumed that the development of what physicists call *classical physics* is a hinge of history, a great turning point that we call *the rise of modern science*, a phenomenon we associate with the seventeenth century. Modern *physics*, at least to physicists, is what happened in the twentieth century. Forgive our confusion on the double usage of the word *modern*, but we must live with both the usage and the confusion. This propagates into considerable confusion in the use of the word *modernity* in theology and the arts, *post-modernity* in literature and the arts, etc.

What does God and theology have to do with the concepts of classical physics which is given such priority in the world-changing scientific revolution? It is certainly true that this *modern science* arose in the Western culture of Christianity, a Christianity recently stirred by the rise of Protestantism. Attempts to give Christian theology, or more specifically the

Protestant version, credit for the rise of science have the feel of a dog chasing its tail. Which is the dog and which is the tail? I fear that many present-day tendencies to argue that Christian theology gave rise to science have implicitly assumed that science is good and supreme and we (as Christians) can somehow justify ourselves by our attachment to something bigger and better than us. An ape can boast that it was what led to humans. A good example of this is "scientific creationism"; creationism, a belief in a God of creation, is presumably bolstered by becoming "scientific" as though science is a higher authority. However they are connected historically, classical physics and classical theology tied to Western Christianity have some strong bonds. Let us explore very briefly a number of these with obvious overlaps and intertwining:

1. God is a God of order. God speaks and it is so. God lays down the law, and Nature obeys. Therefore, we should expect to find lawfulness in Nature.
2. God is faithful and everlasting. Therefore we should expect conservation laws of things that do not get created or destroyed. Being, enduring existence, like that of fundamental atoms or substances, is therefore a primary quality on which we could base our knowledge.
3. God is omniscient; he knows it all. Therefore all is determined and we, created in his image, should be able to learn about the creation, at least partially, maybe by pure reason by which we have been endowed.
4. God gave us a mandate to care for creation. Therefore we have a need, maybe a right, to be able to comprehend it.
5. God is a revealing God who wants us to know and understand so we can have a relationship with him.
6. Creation comes from *One* Creator. Therefore we should expect a unity and simplicity tied to the *One* source. We should also expect a universality to be apparent in the laws and properties in Nature.
7. God is autonomous and therefore makes the world according to his choice, not some *a priori* template which is prior and higher than him. Therefore we have to go to Nature and look to see which of the infinite possibilities God chose. Nature is not necessary, a fixed world that must be this way; it is contingent on God's choice (and maybe our or Nature's choices).
8. God made the world perfect, so that it can run by itself, independently of him.

9. God is the First Cause, the Prime Mover. Therefore, causality, a series of secondary causes is a good model for how the world operates.
10. God loves us and made this universe for us to understand and to use for our benefit.
11. As creatures in God's image, we have (at least partially and imperfectly) the mind of God and therefore should be able to think God's thoughts after him by reading his book of Nature.
12. We are individuals and are individually responsible. Therefore we have a right and responsibility to ferret out truth for ourselves. Just as we, in Protestant tradition at least, do not rely on the priest or church tradition and authority for our revealed truth and salvation, we must go and see for ourselves. The test of truth is to experience it yourself, i.e., do the experiment. Thus we have the emphasis (sometimes forsaken) on laboratory and field experience for our science students. The say-so of teachers and books is not enough. (The motto of the Royal Society is: *Nullius in verba* (Take nobody's word for it.))
13. God is absolute and beyond time and space; he created both time and space and then filled it with matter which can then do its dance. In other words, God made the theater (space), set the performance times (time), provided the cast (matter), wrote the script (laws of motion), raised the curtain ("Let there be . . ."), and shoved the dancers onto the stage with a carefully planned choreography to follow.

The result is our prototypical science, God's story written under the *nom de plume* Isaac Newton (1642–1727) and known as classical mechanics. (We might admit under questioning that Newton himself is not God. Alexander Pope said: "God said, 'Let there be Newton,' and there was light." OK, make it God's story as told by Isaac Newton.) Thus the world can be reduced to three fundamental measurements: length (measurement of space, L), time (temporal measurement, T), and matter (measurement of mass, M): What? [M] Where? [L] and When? [T]. One law linked these together *causally*, using a concept called *force* as our answer to: Why? A complete physical description consists of finding all the pieces (the *whats*), all the forces (the *whys*), and using the law, Newton's second law, to follow *where* and *when* the pieces go. This is still the basic agenda of physics, summarized by a fundamental set of particles (like the elements of the periodic table, but smaller in number) and fundamental forces of interaction. All physical concepts can be reduced to simple combinations of length [L], mass [M], and time [T]. For examples: velocity = [L/T], acceleration = $[L/T^2]$, momentum = [ML/T], force = $[ML/T^2]$, energy = $[ML^2/T^2]$, pressure = $[M/LT^2]$, volume

= $[L^3]$, action = $[ML^2/T]$, etc. Beautiful simple theology resulted in beautiful simple physics. The world *is* and the world is *knowable*. To say it in big words of philosophy designed to sound good: ontology and epistemology are one. The being and becoming of the universe are knowable in a basic law about very fundamental givens: space, time, and matter. This Newtonian worldview is satisfyingly simple, a sparse sketch of all there is and all that happens. It is also very productive. The exploitation of nature is very straightforward. Newtonian mechanics is sufficient for amazing things: as Kenneth Boulding said, going to the moon was the crowning achievement of the seventeenth century.

If you have been reading between the lines, or maybe reading previous chapters, or have been approaching some of our theology warily, you may know what our response should be to this epochal hinge of history, this first and only course in physics for most people who are "privileged" to study physics at all: tommy-rot, hogwash, nonsense, wrong way. It is basically wrong and we have to sort out our theology and our science, and see if we can figure out the places where we can easily take wrong turns. Let's be clear that there are lots of good theological principles here and the result has lots of good physics, but the worldview we end up with is really screwed up.

If you can see where we went wrong, you are better than I and better than a host of well-motivated, well-disciplined, well-educated genius Christians who got us into this mess. Let us see, if we can, how this all came together.

We may certainly hearken back to some pre-Christian thinkers, but let me go back to two of the early scientific practitioners, both well-connected in their Christian faith: Kepler[1] (1571–1630) the Protestant and Galileo (1564–1642) the Catholic. These are not the only giants, but they are two of the most important giants on whose shoulders Newton stood to see farther than others. Both were deeply involved in their respective churches and thus in the troubles and opportunities of the struggle between the churches. Both forged the middle path which combined heaven and earth, mathematics and physics, the theoretical and experimental, the empirical *a posteriori* bottom-up approach and the heady symbolic *a priori* top-down approach. It is hard to imagine two people who were more deeply caught in the vises and vices of religious differences and in the struggle between religion and a newly emerging science.

Kepler was rescued by the Lutheran church from poverty and difficult family situations and educated with the goal of the Lutheran priesthood.

1. You may want to read the following: Ferguson, *Tycho and Kepler;* Kepler and Baumgardt, *Johannes Kepler;* Caspar, *Kepler;* and Koestler, *Watershed.*

Although the new Copernican solar system (published 1543) with the Earth and other planets going around the sun was not officially taught, Kepler learned of it through a teacher and was taken by it for *theological* reasons. We were not the center of the universe and we revolved around a higher power in perfect circles. No one could be more Platonic and mystical in his beliefs in the perfection and symbolism of spheres. He spent years with models of nested Platonic regular polyhedra and spheres as the basis for the solar system; even the number of planets was clear to him by the mystery of the meaning of numbers. The ideality and supposed ultimate reality of Forms and the magic of numbers, both prevalent in the Platonic/Pythagorean worldview, threatened to put Kepler into a paradisiacal fairyland which would explain the world from pure mathematical/theological reasoning, or, if you like, revelation. The reality of theological conflict, where his beliefs threatened his livelihood, his life, and his standing in the church, and the need to make a living by teaching and astrological readings kept him grounded in terribly earthy matters, including the trial of his mother as a witch. The Lutheran church may have educated him, but the threat of poverty and family problems never went away. Astrology depended on data of the heavens, but also on such matters as the exact time of conception. (He noted his own as 4:37 a.m., May 17, 1571, with his parents' marriage on May 15 and his birth on December 27. I'll let you do the math and consider what some family problems might have been. He was a sickly child, perhaps blamed on being a "seven-month baby.")

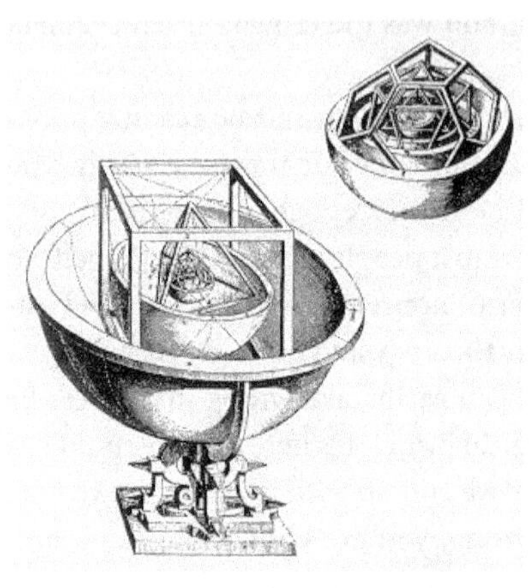

Kepler's elaborate scheme for the solar system, nesting alternating spheres and regular polyhedra, pictured in his *Mysterium Cosmographicum* (1596)

The premier astronomical observer of that pre-telescopic era, Tycho Brahe (1546–1601), finally ran afoul of the Danish King Christian IV and left his Danish island of Hven, taking his observatory, his expertise, his data, and his quarrelsome spirit to a more southern kingdom in Prague. His former subjects, glad to be free of him, immediately dismantled the nobleman's castle for their own use. Kepler wrangled a position with him for the last year or so of Brahe's life. The gold-mine of data fell to Kepler (or he stole it) in one of the most fortunate events in scientific history. Kepler mined the data with incredible patience, perseverance, and insight for years. The Copernican system, accepted for theological reasons, proved its worth in the careful mapping of the planetary data into mathematical laws of motion. The spheres, so dear to his Platonic theological thinking and the circular orbits which went with them, had to yield to elliptical orbits, lacking in the symbolism but equally beautiful in the mathematics. Moreover he formulated a law for the varying speeds as a planet traversed the ellipses, slowing at large distances from the sun and speeding up close to the sun. And the total time for an orbit, the year in Earth's case, was simply related to the size of the orbit, large orbits being slower than the smaller ones. Starting from the belief in God's use of beautiful symbolic forms to lay out the solar system, he developed an equally beautiful mathematical set of rules which came from hard work with hard data, an unprecedented marriage of looking hard with the outer eye and keeping faith and focus with an inner eye stayed on God, perfection, and beauty. He is a joy to read because you can feel both of these guiding lights.[2]

Kepler also brought another aspect to the astronomical science. The perfect symbolic circular orbits were assumed to require no explanation; celestial spheres were self-moving by virtue of their divine nature just as Aristotelian motion was due to the nature of things (e.g., heavy things fall to Earth, their home). The varied mathematical relations of planetary motion begged for a causal explanation and Kepler tried, a veritable scientific revolution in the very attempt. His explanation was unsuccessful by our standards, but the idea that the sun somehow should and could sweep the planets around was a precursor to the Newtonian gravitational force which kept them from flying away.

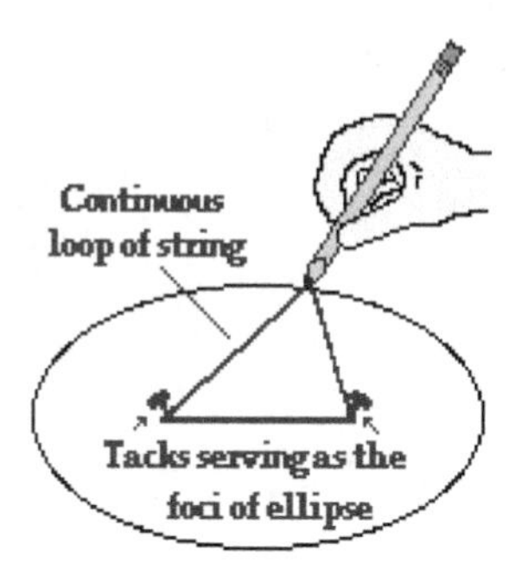

Kepler's first law: Planets travel in ellipses (pencil point) around the sun at one of the foci (tack position).

2. See, for example, Kepler, *Six-Cornered Snowflake.*

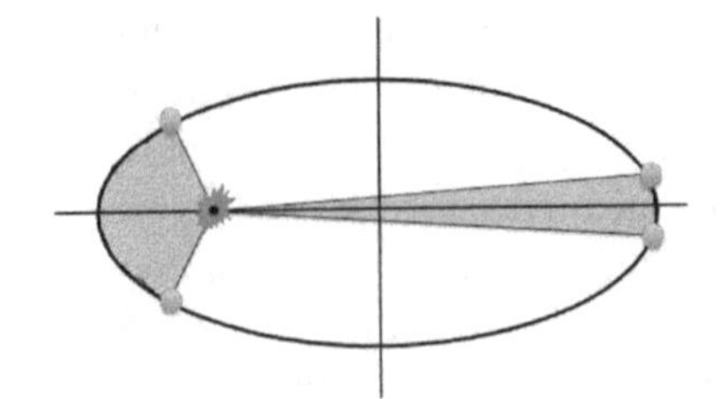

Kepler's second law: The line from the sun to the planet sweeps out equal areas in equal times, the planet moving faster when close and slower when far from the sun.

$P^2 = ka^3$

Kepler's third law given in an equation relating the period P (the year in the case of the Earth) and the average distance of the planet from the sun, a. k is a constant, the same for all planets.

Galileo's story is much more widely known, if also much more wildly distorted by the conflict between science and religion motif. He managed to bring the celestial down to earth and also elevated the mundane to ethereal realms. While churchmen literally refused to look through his telescopes,[3] he saw that the heavens were quite imperfect and earth-like: mountains on the moon, Jupiter's own copy of a solar system, phases of planets like phases of the moon, etc. Looking at motion in our imperfect world, he saw past the imperfections of air resistance and friction to beautiful mathematical relationships for falling or projected objects. A pendulum swing and the time of fall were independent of mass (*ideally*, in heaven as we think it should on Earth). The pendulum clicked off time independently of the extent of the swing; it did not slow down (in time) as it lost its energy. The falling body or one rolling down a smooth incline plane from rest clicked off distances of 1, 3, 5, 7, . . . (for any units) like the even ticking of the pendulum clock. The beautiful simple rule that the odd numbers add up to the squares of integers gave the ideal motion a simple quadratic equation. $1 = 1 = (1)^2$, $1 + 3 = 4 = (2)^2$, $1 + 3 + 5 = 9 = (3)^2$, $1 + 3 + 5 + 7 = 16 = (4)^2$, etc.

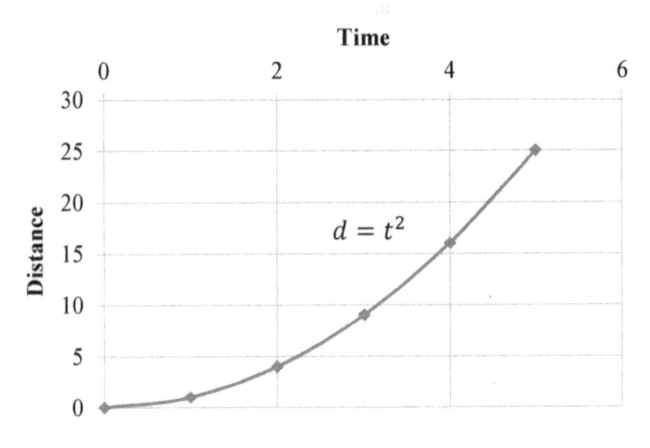

3. Koestler, *Watershed*, 190.

The ability to see past the imperfections of our experiments to a pure and simple mathematics started the cottage industry we joke about in our unrealistic idealistic physics courses: the market for massless springs, frictionless inclined planes, massless pulleys, massless ropes, etc. One is tempted to call them Baptists, where they never have mass, but I suppose not many Baptist churches are frictionless. The ability to see through to the ideal, even when we live in a clearly non-ideal world, was science copying from religion's notebook, however rightly or wrongly we might do this in our religion or our science.

Philosophers had set up this two-sided world from the beginning of philosophy and continue it through the scientific revolution up to today. Plato (c. 427–348 BC), to whom all (Western) philosophy is a series of footnotes (according to Whitehead), accorded ultimate reality only to the world of ideas or Forms. He was deeply affected by one of his heroes, Pythagoras (c. 570–c. 495 BC), and accorded mathematical truth and symbolism the highest respect. Mathematics was the prerequisite for entrance to his university, the first in Western civilization, and we humans, somehow able to tap into, remember, or just know that truth, were trapped in the unworthy bodies of a fleshly material world which represented evil and from which we hoped to escape. While Socrates willingly accepted death as an escape from this mortal life, his pupil, Plato (and his followers' teachings which were appropriated into Christianity) eventually taught us of immortality of the soul and its escape to a literal heaven. Heaven, the heights, the celestial sphere is an understandable metaphor for a world of an afterlife, of an eternal life, as the abode of God who is above all. An evangelical Christian will be concerned about your future in heaven or hell. Read the Old Testament and New Testament carefully for the language. The Old Testament basically has none of that "everlasting" language. The New Testament is really speaking of the Kingdom of God and struggles with the language, Greek, in which the New Testament writers wrote. When John promises everlasting life in John 3:16, the context is one of being born again, being born of the Spirit, a newness, a different quality. It is a passage of escape only in the sense that birth is an escape. The interlinear Greek New Testament I consulted leaves the Greek word which we translate as *everlasting* essentially untranslated, using a transliteration *eonian* or *aeonian*, which we never use but has the clear connection to *eon* which we do use (at least in crossword puzzles). When we try to describe something outside of our world, we have to rely on *timeless, infinite, with no beginning, with no end*, etc., but we often cannot envision another *quality* of world, a Spirit-world, a God-world into which we need a new birth. *Heaven*, as we find it in the New Testament, translates the Greek word *ouranos*, which betrays its Greek mythological roots by our naming of the planet Uranus, the god of the sky. Plato and his teacher Socrates helped us into the concepts of a sterile otherworldly ideality better associated with mathematics than with a relationship with a person, the king of creation. This world of ideal mathematics is still clearly the kingdom of God which theoretical physicists see. This is the impersonal God of Einstein, a God of perfect law and order, summarizable (he hoped) in a *unified field theory* or a *theory of everything*. Such equations, such theories, can be beautiful and symmetrical and consistent, but to be the truth, they have to get up and fly, as the universe does. Equations are seductive, but they do not have life.

The god of Aristotle (384–322 BC) may be as sterile, but it escapes the other-worldly mathematical feel. Aristotle came from a medical family, and he stayed to a large extent grounded in that world of life, biology. Darwin was awe-struck when he discovered Aristotle's work.[4] Aristotle described some biological phenomena which were only rediscovered many centuries later, ignored in his writings because no one else had seen them. Aristotle's logic was important for the deductive world of mathematics, but the reverence and reality accorded to Platonic geometric forms and Pythagorean number mysticism were missing. Aristotle's view of the universe was organismic, founded in life. He got his hands dirty, both in collecting samples and dissecting them. Rutherford might have called him a stamp-collector who could not do the mathematics, could not see through to the ideal mathematical world beyond our very impure world. But Aristotle was deeply involved in seeing the real world, i.e., the visible world, the world of our senses. He could see that a ball which was rolled would come to rest. To teach classical physics, we try to break through that Aristotelian reality thinking: an *ideal* ball (one with no frictional losses) would continue to roll. Aristotle did not see the unseen that we "see" or invent. When we see circular motion, we may see three possibilities: 1) it is natural and requires no explanation; 2) its path is natural, but it will stop if something does not keep pushing it around; 3) the natural motion is constant speed in a *straight line* and it will curve around in a circle only if the right force pushes or pulls it toward the center. Aristotle saw the first two: the celestial spheres were related to the Unmoved Mover and had the perpetual movement and form built in, and the lower spheres were basically pulled around by the Prime Mover. Again we see an impersonal God but one at least linked to maintaining the motion. Feynman (1918–88) said that the progress in physics over the 2,000 years from Plato/Aristotle to Newton consisted of moving the angels who provided the

4. Darwin: "Linnaeus and Cuvier have been my two gods . . . but they were mere schoolboys to old Aristotle." Jaki, *Relevance of Physics*, 30.

forces for circular motion from the backside (pushing it along) to the outside (pushing it inward to the center).

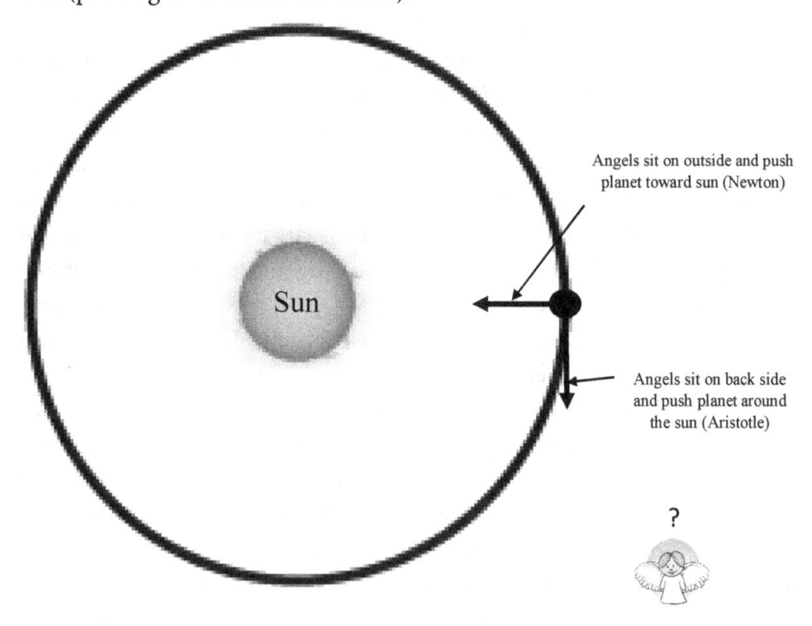

The primary philosophers of the beginnings of the scientific revolution in the seventeenth century were Francis Bacon (1561–1620) and René Descartes (1596–1650). Francis Bacon took the mantle of Aristotle while Descartes carried the baton for Plato.

Francis Bacon set the tone for our junior-high characterization of science: observe and experiment. Torture nature with your questioning and Nature will yield its secrets, like a mine yields its ores. Nature was made for mankind, so we should go take it, waterboarding if necessary. The method of science was as mechanical and sure-fire as the classical picture of Nature became. There are no secrets, no invisible back-story to worry about; just keep digging and accumulating data and experience. Mathematics has no special place, but Nature is sure to tell the whole truth, and nothing but the truth, as if the laws were carved in stone.

Descartes is famous for fighting through the uncertainties which his extreme doubt forced on him to the one certainty: "I think, therefore I am." *Cogito, ergo sum*. It probably should have been: *Dubito, ergo sum*. "I doubt, therefore I am." But his doubt gave him certainty of himself and it ended in the same Platonic ideals, the Forms which are eternal and somehow known to the mind. The most important result of this approach was a dualism which separated mind, a non-spatial non-material thinking thing, *res cogitans,* from the material world, *res extensa,* basically things that take up space. If one's mind, the thinking doubting thing, was the origin of certainty and reality, the unreal or dubious material world managed to become the machine, the real-life carrier of the ghost/mind. You can easily see the natural progression to a mechanical universe in which God is the Ghost in the Machine. Descartes, in spite of our characterization of him as a mathematician and philosopher, was very interested in biology as well, but his biology worked like a machine. He sought eternal life for that machine by thinking the machine could be perfected. In a wonderfully ironic statement, he said he could fix the machine, our life, to last forever if only he could live long enough. Physicists find his physics forgettable, but he made major mathematical progress which contributed to the mathematization of physics, eventually giving physics (in Newton's formulation) the certainty and precision we could only envision in pure mathematics.

Newton had both the right timing and the mathematical genius to take the empirically measured world, especially Brahe's measurements of planets as digested into Kepler's laws, invent a mathematics of certainty, find a mathematical form for gravity, write a law of motion for objects in the God-given absolute space and time, and thus construct the sufficient Skeleton for the complete Machine/Universe which the Ghost/God could inhabit. It is said that a skeleton cannot walk, but one cannot walk without a skeleton.[5] This Newtonian Skeleton seemed so well-made that it could walk without God. Well, God may have made it and started it walking but a declaration of independence was quite natural, the beginnings of deism and most of our modern atheism. The power of mathematics was overpoweringly awesome. First, Newton created calculus. Then he wrote a law of motion, telling how forces change velocity. Then he saw a simple mathematical formula for the gravitational force of mutual attraction of all masses. Then, quite literally, a few pages of algebraic manipulation, perhaps squeezed onto a single page, and out pops the results of two genius lifetimes, Brahe's astronomical measurements and Kepler's analysis of them. No wonder that Newton and his achievements were lionized. Every major thinker in every field for several centuries thereafter used Newton as a model and a standard. If Christianity struggled to incorporate, digest, or spit out the Greek philosophy during its origins, we must be aware that our present faith under which we as Americans presume our nation to be conceived and developed must have the same concerns with the Newtonian scientific revolution. Did Newtonianism come unerringly from a Christian worldview or do we stand the risk of being more Newtonian or even more American than we are Christian?

5. Jaki, *Road of Science*, 330.

At least for me, it's not an easy question and it is not to be ignored, even if I cannot answer it.

The result of the Newtonian scientific revolution was the completeness of a perfectly accurate accounting and knowability of the movement of every piece of mutually exclusive, eternally lasting, space-filling matter. In other words, the world is made of things which move according to perfect laws. The things are forever and move in a provided arena of space and time according to perfectly causal laws. This has been the model and the framework of all thinking in the three centuries following the seventeenth century. It is the conceptual basis of modern thinking that is totally bankrupt; none of the concepts hold up to careful scrutiny.

If one can appreciate the beauty and form of the simple (to a mathematician) mathematical structure of the geometry, analytical geometry (geometry done in formulas), and calculus, one can get sucked into the Platonic reduction of the universe. Aristotle, the rebel against mathematics can provide some corrective. His account of causality is so much richer that it allows a larger framework. He identified four causes: 1) material cause, 2) formal cause, 3) efficient cause, and 4) final cause. The block of stone was the material cause; the formal cause was the shape formed in the sculptor's mind; the efficient cause was the hammer and chisel which chip away the stone; the final cause was the overall purpose, the image and the display in honor of someone or beautification of the public square. The Newtonian causality is reduced to the kickable matter, which is trivialized and taken for granted because of its givenness and apparent indestructability, and the visible chipping of stone. The important causalities of purpose, planning, imagination, and design are ignored. The Newtonian causality represents a mere skeletal shadow of the full-bodied organismic purposeful causality of an Aristotelian whole.

A good philosophical challenge would rip apart the pieces of such an overly simple view of nature. Fortunately physics with the advent of relativity, quantum physics, and other elements which we call modern physics has also ripped such a narrow view apart. That is why physicists of the twentieth century were sometimes given credit for being philosophers and could get by with suggesting physics was a prerequisite for philosophy. Even as pre-twentieth-century thinking developed the superstructure of classical physics, it exposed the cracks which later widened and deepened into chasms. We glossed over the cracks in our thinking as we went, but in hindsight we can see that we should have seen our own weaknesses. The next three chapters will show some of what was invisible and papered over.

4

This Little Mine of Light

> Light has symbolically illuminated religious thought for millennia, and proved subtle enough to shake the Newtonian particulate view.

WHEN WE THINK OF the world as made of moving masses, impenetrable hard objects which were and are and evermore shall be, we close our eyes to the "thing" which allows us to see, light. Nothing has been more essential to life as we know it, and nothing has been more symbolic of divinity, of beginnings, and of mystery. The Judeo-Christian Bible starts with "Let there be light . . ." and one would be hard-pressed to find a religious urge which does not have a god of light or does not consider the light as image for knowledge and righteousness. The spatial image of God and heaven high above is not accidentally related to the greater and lesser lights being found in the heavens, high above. And I think that no part of the physical world has been more mysterious than light; yet it is key to understanding Nature and holds great promise to unlock other mysteries. As it illuminates the rest of the world, it exposes our ignorance. While *Enlightenment* is used as a description of spiritual heights and superior knowledge, light has been the most elusive of the physical concepts. While it seems to harbor one of the absolutes of nature (the speed limit of the universe), its essence seems to be more creative than our imagination. It seems to be as complex and pervasive as God or the universe, but it can also be corralled to respond to the binary yes-no questions, yielding the smallest piece of information, the bit, which is at the heart of digital computing. The tale of our evolving understanding of light is central to quantum physics and to all physics. It deserves a book of its own (and of course has inspired many). One or two chapters will not contain it. We will be hard pressed to find any area of physics which has not felt the penetrating gaze of optics or its gentle illumination.

For most of mankind's existence, light has been enchanted; the first creation of God in the Judeo-Christian story and the gift of the Egyptian Sun-god Ra are just two examples. Ra: "I am the one who openeth his eyes, and there is light; when his eyes close, darkness falleth."[1] At the hands of the medievalists, light had its elaborate metaphysical dress of many cultures stripped from it and the seventeenth century, especially Descartes and Newton, flayed its flesh, leaving a skeleton in the image of material mechanics. In the metaphor of Fontenelle, secretary to the Paris Academy of Sciences in 1686, Nature *may* be seen as a dramatic passionate meaningful production but was seen by science for what it *really* is, the technical backstage mechanical operations of the stage engineer and stage-hands, make-up artists, etc. This is the dream of the age of the mechanical cosmos and the nightmare of those of us who want to see the dramatic production. If the latter includes you, don't get too worried: Dem bones gonna rise again, flesh and all!

Descartes (1596–1650) may have been the dwarf who built a gigantic foundation of a mechanical universe based on reason, but Newton (1642–1727) climbed on his shoulders to erect the structure. Newton's theory of optics, published in *Opticks* in 1704, rode the wave of his success in mechanics, gravitation, and astronomy drowning all other concepts in its wake. Optics was like mechanics, simply a movement of *corpuscles*, particles of light. Reflection was like billiard balls bouncing off a barrier, preserving their motion parallel to the wall and reversing the perpendicular motion. Refraction, the bending upon transmission, was no harder to explain: the parallel component stayed the same and the perpendicular component was lessened (it slowed down) or increased (it speeded up).

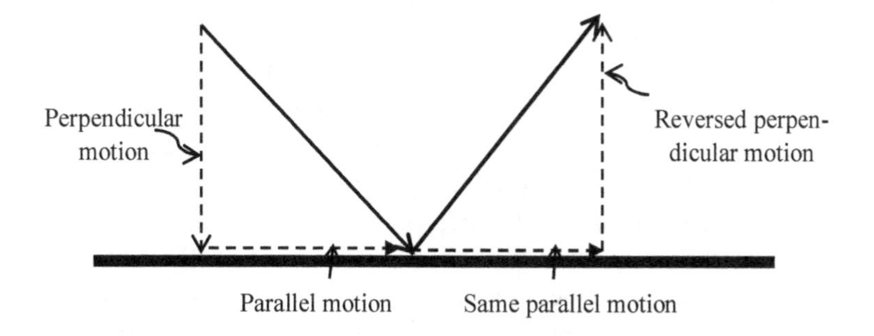

1. Zayonc, *Catching the Light*, ix.

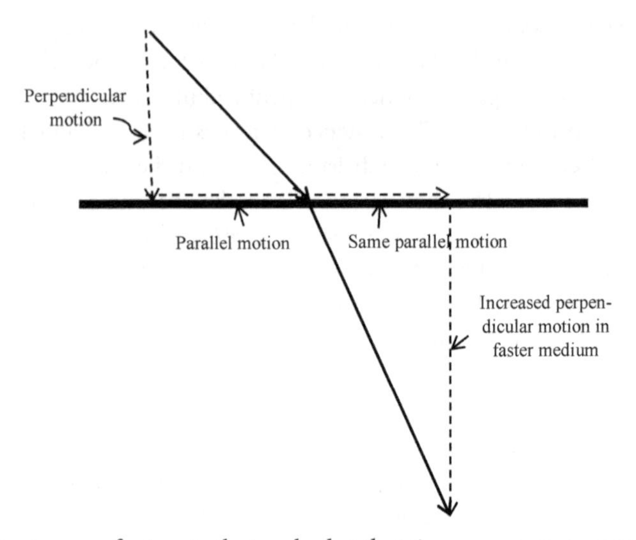

Newton was fortunate that we had at that time no measurements of the speed of light in various materials, because his theory ends up being dead backwards; the picture above is the way light bends in going to a denser (slower) medium. Newton was also quite aware of what we call interference phenomena; patterns which we call Newton's rings were used to test the curvature of lenses. His "explanations" involve "fits of reflection and refraction" and should convince no one. Don't get me wrong. Newton did great work in optics, but he also missed the mark by miles. And, as we will see, our twentieth century hero, Einstein, was able to revive a particle theory of light. The games of science are never over till they're over (as Yogi Berra says) and you should not bet on them (as Pete Rose should have said).

Christiaan Huygens (1629–95), an elder contemporary of Newton, had a different vision of how light travelled. In spite of Newton's overpowering influence for the century following, Huygens's view of optics had more staying power than Newton's. Light a la Huygens was a wave spread across a wave front like a row of marching soldiers maintaining a line (not necessarily straight) as they move to the position of the row ahead, producing a series of wave fronts or rows of soldiers. The lines get changed in shape or direction by two means: a reflecting barrier or changes in speed or step size. Just as soldiers (and marching bands, for us non-military folk) keep time even when they move at different speeds, waves of physics do as well, maintaining the tempo (beats/minute) or frequency (cycles per second, now named Hertz (Hz) in honor of a nineteenth century physicist) and varying step size (wavelength) to change speed. Reflection, refraction, and shaping the wave front for focusing can be demonstrated in the following pictures.

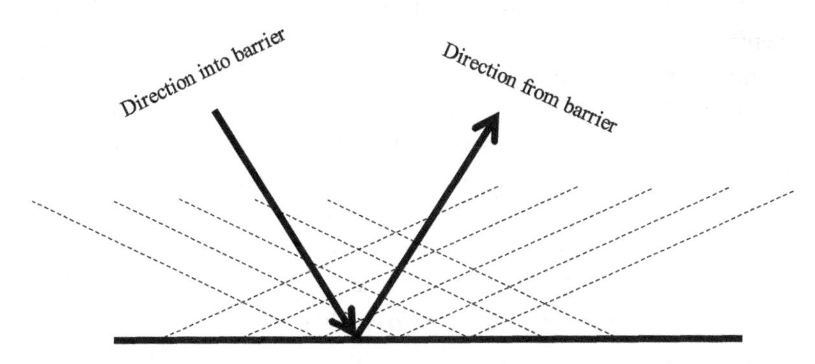
Direction into barrier
Direction from barrier
Reflecting barrier

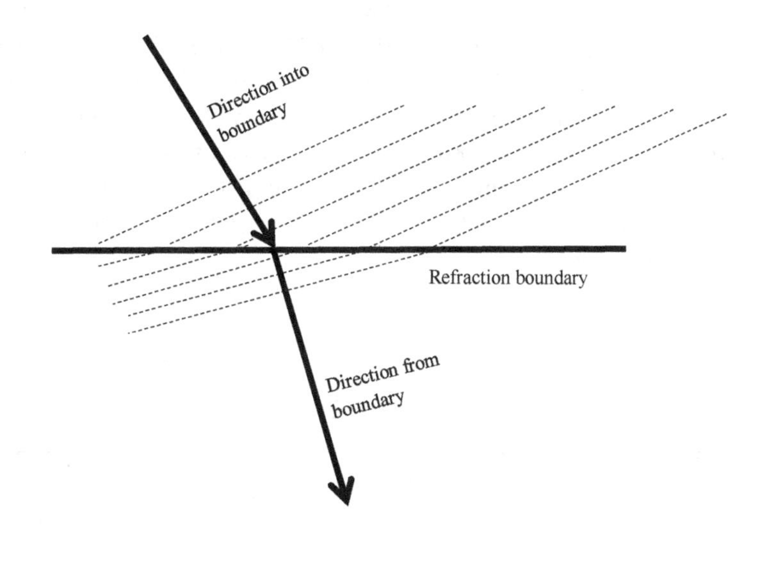
Direction into boundary
Refraction boundary
Direction from boundary

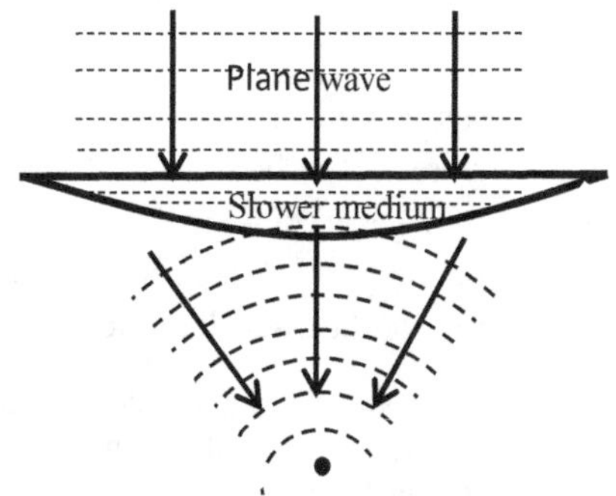
Plane wave
Slower medium
Focal point

Another elder contemporary, Pierre Fermat (1601–65), a mathematician, had a quite different model for following the path of light travel, called the principle of least time (more generally, an extremum principle, giving a minimum or maximum, depending on circumstances). This is an idea which has often excited great interest because it is global and more than hints at the involvement of a designer (e.g., Leibniz's belief that the world is the best of all possible worlds, in spite of its all-too-clear flaws). This idea actually goes back to Aristotle, a concept he called *economy of nature*. One of its most devoted and vocal followers was Maupertuis, an interesting character.[2] He introduced it in 1747 into mechanics, albeit in a somewhat defective form, with the grand claim as "an exemplification of the perfect wisdom of God."[3] Max Planck (1858–1947), whom we will meet later, was a firm believer and Feynman (1918–88), one of the twentieth century's greatest theoreticians, carried it as a lifetime motivation, learning its version in mechanics (the principle of least action) in high school. The idea that light (or any particle) can check out all the possible paths and choose the *special* one is indeed intriguing. Below we show two examples: reflection and refraction. The very special condition for *focusing* is that *all* paths between two points take the *same* time. Note also the method of drawing ellipses shown earlier with Kepler's first law and you will see why the two special points are called *foci* or *focuses*.

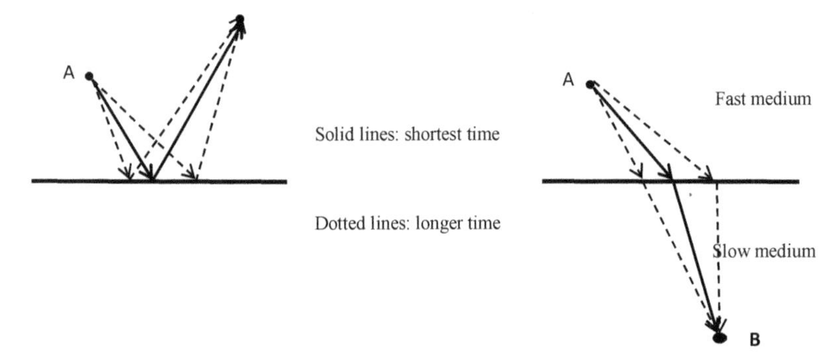

I think I was correct in stating light's metaphysical dominance across all cultures, but I do need to backtrack a bit because the Judeo-Christian tradition does introduce a stronger magic, a prior attribute of God: the Word, i.e., sound. "And God *said*..." comes before "Let there be light." Sound is almost as mysterious as light, but it fell prey to the disenchantment brought by scientific

2. He is described as a musketeer, biologist, moralist, linguist, metaphysician, and physicist in: Jaki, *Relevance of Physics*, 65. For a full treatment of this extraordinary man, see Terrall, *Man who Flattened the Earth*.

3. Quoted in Margenau, *Nature of Physical Reality*, 186.

explanation a little easier than light. Reducing music to wiggles in some musical instrument which wiggles the air which wiggles something in your ear is the same travesty as reducing Michelangelo's *David* to marble dust, but it is the stuff of modern science. Even though we know that music and the sculpture is *more*, it does set a mind-frame that allows scientists today to reduce religion to spots of activity in brain-scans. Nature does take its revenge for the inquisition we impose upon it. Since light and sound have similar behaviors, we get a hint of what light can be by our stripped-down model of what sound is.

Almost exactly 100 years after Newton's *Opticks* (1704), Thomas Young (1773–1829) challenged the Newtonian corpuscular theory by a principle of addition which just does not seem to work for particles. Addition had spawned its opposite, subtraction, at least as far back as Job, the oldest (I'm told) book of the Bible: "The Lord giveth and the Lord taketh away. Blessed be the name of the Lord." But we had learned to treat subtraction as the addition of negative numbers, x—y = x + (-y). We know how to add particles and take particles away, but how could you even conceive of a negative particle added to a particle to produce nothing, zero, nada, zilch? *Material things* just don't behave that way, and Newton's corpuscles were *material things*. To suggest that light could occupy the same place as other light and that light could either add *or* subtract required genius, daring, and arrogance, not to mention some visible experimental results. But water waves and sound waves cross in space and the intersecting waves may enhance or cancel each other. We call this addition/subtraction the principle of superposition and the phenomenon produced we call interference. An ambient level, the height of the water or the pressure of the atmosphere without a wave, represents a *zero* from which changes can be up or down, plus or minus. Why could not light be such an undulation of a universal luminiferous ether? What an irony, what insanity, that light, our source of seeing, should be supposed to be made up of *invisible* wiggles in an *invisible* medium. Such talk can make the metaphysical God-talk sound really down to earth. Nowadays you can go to the store and buy noise-reduction head-sets which *add* sound (negatively) to the ambient sound to cancel it, but we don't (yet) have light additives to cancel what we see. It was not easy to do optical interference experiments in 1803 but you can easily perform Young's experiments today. A laser (pointers used by lecturers) shining through a small slit (place two strips of opaque tape next to each other on glass) will give you a blob of light on the wall behind. Produce two slits very close to each other and you should get a blob plus a blob, which just looks like a brighter blob. But it doesn't happen that way. The blobs *interfere* and produce stripes of bright lines and weak lines (maybe even *no* light), which we call respectively constructive interference (enhancement) and destructive interference (cancellation or reduction). (This is also not easy to produce with sound because pure sounds, single frequencies, are uncommon, and reflections and two detectors (ears) confuse the issue.) With LEDs (light emitting diodes) being ubiquitous, one can see the effects easily by just looking through two slits at an LED. Here is a picture of laser light taken through two adjacent slits, called Young's two-slit interference pattern.

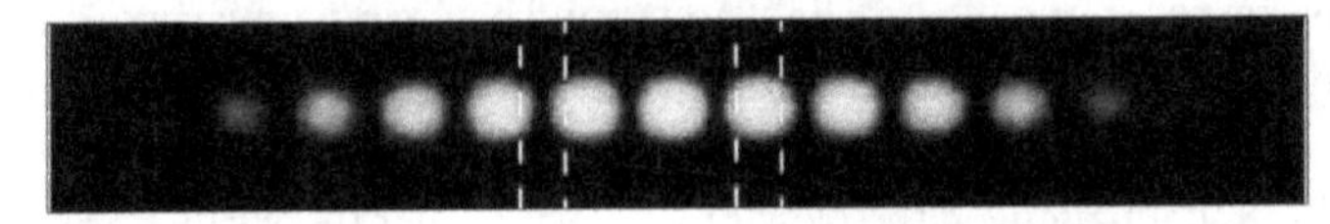

Not much later (1819), a French construction engineer, Augustin Fresnel (1788–1827), similarly dared to enter the lions' den of corpuscularists with a wave theory of light.[4] The Paris Academy of the Sciences offered a prize for the best paper submitted on the problem of diffraction, the bending of light as it passed around barriers or through apertures. Experiments were difficult but Fresnel produced a sophisticated mathematical wave theory which matched the difficulty if not the results of experiments. The committee consisted of such a stellar array that their names, Biot (1774–1862), Poisson (1781–1840), Gay-Lussac (1778–1850), and Laplace (1749–1827), are still well-known to scientists two centuries later. They were a formidable lot and they were antagonistic to a wave theory; they still supported the Newtonian corpuscular model. Some of them were also better at math than Fresnel. Thus Poisson solved the problem (using Fresnel's theory) for the diffraction around a small circular obstacle, a problem which Fresnel had not been able to solve. The result was more insane than Thomas Young's patterns (still unknown to Fresnel); a bright light shining on a BB, for example, would of course produce a circular shadow, but Fresnel's theory with Poisson's solution produced a bright spot in the middle of the shadow. Everyone knew this was absurd! Absurd it may have been, but here is a picture. Arago (1786–1836), a friend and supporter of Fresnel, did the experiment and found the spot, changing the insanity into a prize-winning paper.

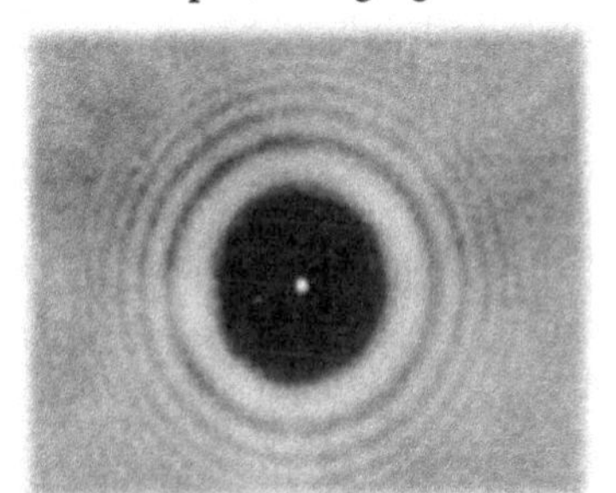

Poisson's spot (more properly but more rarely called Arago's spot).

4. Hecht, *Optics*, 443–4.

Fresnel and Young managed to solve another strangeness of light, what we call polarization. Light appeared to be divisible into two pieces. If you lay a piece of calcite crystal on a printed page, you will see two images, slightly separated. Turning the crystal around causes one image to go around the other. Waves on a string are transverse; the wave travels along the string while parts of the string may be going *up and down* or *side to side.* Thus two independent transverse waves may be propagated together in a medium. They assumed that light was thus transverse. If a filter could be made to allow one polarization to pass while stopping the other, we could see the two different images separately. *And it works.*

Another application comes because the glare from reflections are often mostly of one polarization; a polarizing filter can almost eliminate the glare. The picture on the left is obscured by glare from reflections on the water surface in the top half of the picture. The picture on the right is taken with a polarizing filter which cuts the glare. Both the principle of superposition

and the division of light into two pieces will figure prominently in quantum theory and experiment.

Fresnel and Young gave great impetus to a wave theory, but invisible wiggles of an invisible medium are tough sells. Physics can sometimes make the difficult seem easy and sometimes allows abstract mathematics to be more important than kickable substance or real data, but this kind of magic required several geniuses and a good fraction of a century to sell. That's the subject of our next chapter. But let's take a harder look at the requirements of this supposed luminiferous ether, the supposed medium.

The power of the simple mechanical model of the world with its three elements of mass [M], length [L] and time [T] becomes apparent when we try a simple analysis of this invisible medium which propagates light. We can get a reasonable idea of how fast waves travel by simple thought experiments with waves on a string. There are two competing effects: the strength of neighborly interaction caused by the tension in the string and the inertia of the string's mass. Larger tension (stretching tighter) and lighter string will each increase velocity. Tension is a force (given as $[ML/T^2]$ by Newton's second law); dividing it by [M/L] (mass per length) makes it $[ML/T^2]/[M/L] = [L^2/T^2]$ which is a velocity [L/T] squared. Thus a good guess for the formula for the velocity of a wave on a string is v = square root of [(tension)/(mass/length)]. We are limited by the strength of materials; a fine thread will only handle a certain amount of tension before breaking. The formula changes very little as we go to the three-dimensional world; it still involves the square root of a ratio of a measure of neighborly interactions (how hard neighbors push or pull on each other) to the mass (actually mass per volume, which we call density). Light travels incredibly fast, roughly one million times faster than mechanical waves in our ordinary media. With the square root in there, this requires a medium that we could reasonably call the luminiferous ether to be one-millionth as dense as ordinary matter but one million times the pressure or stiffness which measures how strongly neighbors interact. One million times lighter (in weight) but one million times stronger! That is some tall order, some super-medium, almost impossible to believe. But believe they did; what else could they do? The fact of light may not be kickable but it is incontrovertible. However extreme the explanation might have to stretch, we believe it *if* we believe everything has a cause, an explanation. This is where religion and science appear to be inseparable. The facts of life, the physical world, the spiritual world, the psychic world, the moral world are unavoidable to those who are willing to look hard at them. Physicists looked with sufficient care at light's properties and they believed in an unbelievable ether. Darwin looked with sufficient care at the variety and apparent history of life forms and believed in an unbelievable mechanism and time frame. People of faith are trying to see the whole range of phenomena we experience and believe in an unbelievable God, a person, because our range of phenomena includes more than machines, more than matter in motion, even more than life forms, etc. It has to include consciousness, mind, beauty, love, moral law, social interactions, on and on. However poorly we do it, we need to try to see the unseen and our ritualized recognition of our limited vision or even our blindness is an important part of what we are. A scientist who thinks she sees the end of knowledge or the completion of perfect knowledge is a scientist without hope.[5] A religious person who thinks she knows God perfectly and fully is no longer a religious person. Can one dare to say that life without growth, without a place to go, without a purpose to fulfill is not life?

We may follow the histories of light, ether, and various religious faiths with fear and trepidation. This theory or that theory is past; maybe mine will go away. Einstein supposedly rid us of the ether. But let's not kid ourselves. His general theory of relativity installed another ether, and the new ether has spawned an ongoing expanding series of mysteries of inflationary universes, multiverses, expanding space, accelerating expansion, black holes, dark energy, etc., etc., etc. With regard to light, Einstein said: "All the fifty years of conscious brooding have brought me no closer to the answer. . . . Of course today every rascal thinks he knows the answer, but he is deluding himself."[6] We must, we cannot help but keep "brooding," but if we think we have the answers we are deluding ourselves. All of our great theories, Darwin's, Einstein's, or whoever's, are the *questions*, not the *answers*, and so are our great religions.

5. Margenau, *Open Vistas*, 74. A favorite story for many scientists is in a play by Lessing. Man confronts God, whose left hand holds *eternal search* and the right hand contains *final absolute but static truth*. Which does man choose? "Give me from the left and let the right remain a divine possession." Scientists fancy themselves that way.

6. Zayonc, *Catching the Light*.

5

What Is Reality? Masses or Fields? Ether or Vacuum? Mathematics or Material World? Material World or Spiritual World?

> Faraday's image of lines of force and their subsequent mathematical form by Maxwell as electromagnetic fields, which in turn swallow optics, become a second leg of classical physics, one with clearly different images of reality.

NEWTON HAD SOLVED COMPLETELY the world of mechanics. Well, in principle, he had. We've seen that he really did not get close to solving the problem of light; those Newtonian corpuscles of light, as we've tried treating them in their twentieth-century reincarnation, have no mass. Newton's theory of everything had mass as a central concept. The motion of a mass followed a differential equation. A differential equation is an invention of calculus, the basic idea being that you can get the motion as accurately as you wish, *with no limits*. The basic proofs of calculus involve statements like this: you tell me the accuracy you desire and I can find a way to meet your demands, no exceptions, no limits. We all know there are practical limits to all of our measurements, but the availability of such mathematics gives us a belief in a perfectly accurate, therefore, complete lawfulness. The method of working *all* mechanics problems then is to choose some mass you wish to follow, find all the forces acting on it, plug into Newton's second law, the first Theory of Everything in physics, and turn the crank of mathematics of perfect accuracy, and out comes the complete answer, where and when the mass will be or where and when it was for all time. For the *unreal*istically simple problems which we must pose for *real* physics students, this appears possible and gives us a false sense of certainty and completeness. Let me give you an idea of why the seductive problem cannot be solved, how Newton openly papered over it, and how the extension of Newton's phony solution led us to a whole new realm, which in fact solved the optics problem. Well, partially, for a little while. I've already told you that nobody has really solved the problem of light. But this solution was really convincing and appropriately capped the classical physics picture after a full two centuries of development.

Find all the forces. Students are all too able and willing to invent some pseudo-force to get the answer in the back of the book, so I insist that every force has to have an agent. You have to tell me *who* (i.e., what other part of the universe) is exerting the force. The obvious agents, and the *only obvious* agents, are the things that are touching it, for example, a string tied to it, a floor beneath it, or something butting up against it. Those are easy to name. But then there are the things which are not touching it. That is of course almost limitless, basically every other mass in the universe.

Here is where Newton cheated in a clever reasonable way which was sheer genius and leads us to cheat because we simply cannot do better. Force, hard as it may be for us to believe, was a vague, even mystical, concept which required Newton's law for any clear definition. We could rely on our tactile experience of pushing and pulling for a sense of what forces of contact are. But forces of the distant, non-contact world were (and remain) a mystery rivaling that of light. We call one such force gravity, some sort of force acting between masses through space or some medium, action at a distance we call it. Newton's genius was three-fold: 1) he recognized the universality of the phenomenon (you really have to include *all* other masses if you want to do it right); 2) he gave it a mathematical equation (if you have an equation, you plug in the exact numbers and you get the exact answers); 3) he knew he didn't know what was going on, he did not apologize, and went with what he "knew" (see 1 and 2 above). *Non fingo hypothese*, he said. The loose but reasonably accurate translation: "I won't try faking any speculations." Universality is a double threat from the theological front: God involved himself in the world *and* he made everything the same. Copernicus and Galileo had led us to close the gap between the heavens and the earth, the chasm between the celestial and terrestrial, between Form and matter, which had been part of the Greek heritage. That One Creator should make it *all the same* was a good guess from the Judeo-Christian heritage. What the "it" is and what "all the same" means is harder to come by, and we have good chances to make mistakes. Newton's guess does not even seem close to right from our experience; the sun pulls a planet, a planet pulls its moons, a planet pulls a stone to its surface, but a stone pulls a stone?! No way. Here's where the practical cheating comes in and is justified *in practice* because the pull of a stone for a stone is so small that we can safely ignore it.

Newton's mathematical form is reasonable and comes from a simple idea. Whatever influence a mass can muster must be spread out over all directions, i.e., over the surface of a sphere at whatever distance another mass exists. The force is therefore reasonably decreasing as the area of the spherical surface, $4\pi r^2$, increases. We say that the force is inversely proportional to r^2. Double the distance and the force reduces by 1/4. Thus we see a large reduction of forces for large distances.

But if we consider solid impenetrable spheres and consider the distance between centers to be only a little larger than the sum of radii (they are nearly touching), the spheres have masses proportional to the volume which is $4\pi r^3/3$. Thus the effect for spheres close to each other's surfaces increases with the radius because r^3 increases more than $1/r^2$ decreases. For example, the Earth's radius is thousands of miles while a human-sized or room-sized stone has only a radius of a few feet. Thus the gravitational effect of a ball on a ball in one's laboratory is about one-millionth of what a ball experiences with the big ball we call Earth. You can see that it requires genius or faith or both to see and believe that these two effects are the same thing. And you can see why we can "cheat" for our problems on Earth by neglecting all the masses in the universe except for the big ball, our Earth.

It required an eccentric genius and a wait of over a hundred years (until 1798) for Newton's conjecture about *all* masses to be confirmed and quantitatively measured. Of course, the calculations which matched Kepler's laws made us believe that the sun/planet, the planet/moon, and the Earth/

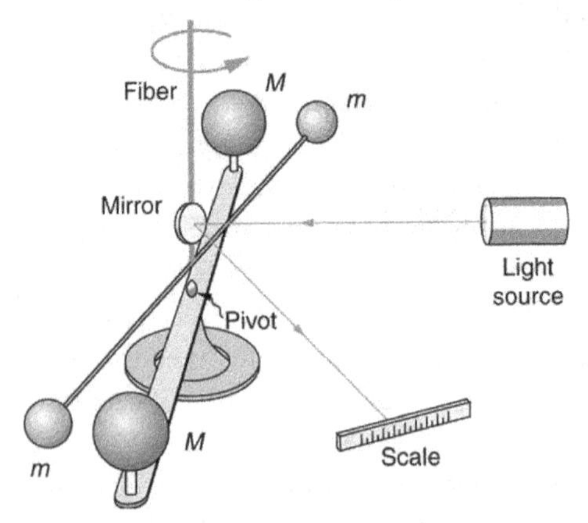

Schematic of Cavendish Apparatus for Measuring the Gravitational Constant G ("Weighing the Earth")

apple-falling-to-the-ground interactions were all the same phenomenon, but it required a clever experiment by Cavendish (1731–1810) (a name familiar to us because Cambridge University named a laboratory after him) to convince us that any and every stone pulls on any and every other stone. By hanging a dumbbell of two lead balls from a fine quartz fiber and measuring the twist in the fiber, such small effects can be measured. To appreciate the care that needs to be taken, you would need to try it yourself. I have, in a basement with concrete floors. All equipment must be turned off, air motion must be shielded, and even taking a single step may cause too much vibration. But it can be done. Because the astronomical numbers, the periods of the moons and the planets were known, this measurement immediately gave the mass of the sun and the earth. Cavendish's paper thus had the pretentious title, "Weighing the Earth." Why he didn't claim to weigh the sun, we don't know.

If you do this experiment, you will realize how weak gravity is in spite of its apparent dominance in our lives on Earth, where fear of heights and falling are legitimate fears. But we also can see that there must be some much larger forces so that the material pieces on Earth (we, for example) do not collapse or deform or form a puddle or thin film. Gravitational forces are totally dominated by forces which give form and maintain form in materials of this world. We now attribute the forces which shape matter to electromagnetic effects. Very weak electric forces (compared to their potential strength) can be produced by rubbing things together, tribo-electricity, we call it. We are all familiar with it by the shocks we get by accidentally rubbing two things together, e.g., shoes and carpet. For light objects, such as strings, balloons, or human hairs, it is easy to have forces which are greater than the Earth's gravity. Everyone has (I hope) stuck a balloon to a wall or ceiling by rubbing it on their hair.

In the century after Newton's prodigious achievements in mechanics and astronomy, the world of electrostatics became something like a circus side-show, parlor tricks with sparks and shocks. They can be very hard to reliably reproduce (little sparks make continuous uncontrollable results) and were not always reliably safe. The story of Benjamin Franklin flying kites and keys in thunderstorms is a good example *not* to try at home (or anywhere else). Nature had been setting up huge electrostatic experiments for all of history. We knew their power and did not know how to control, quantify, or cast them into any lawful equations. But with care and persistence, one can determine that there is some property we call *charge* which we designate as *signed*, coming in a *plus* variety and a *minus* variety. Except for the possibility of both attraction (opposite signs) and repulsion (like signs), we find that it behaves essentially like gravity. We can even do a Cavendish-like experiment and find that the inverse proportionality to r^2 is correct; the interaction ability (which we call force) gets spread to the whole sphere and thus has the exact mathematical form (at least in theory) as the gravitational force. It thus appears that the geometry (in this case the area of a spherical surface) directly translates into a physical law with charges acting on charges in the same way (same mathematical form) that masses act on masses. Both ideas are totally mysterious in their origin but are given quantitatively by the one single simple mathematical form. Because the charge-charge interaction is so huge compared to gravity (based on fundamental particles such as protons, it is 10^{40} = 10,000,000,000,000,000,000,000,000,000,000,000,000 times bigger), clever arguments and clever experiments allow us to be sure that the law is very exact, inversely proportional to $r^{2.000,000,000,000,000,0}$ (see, for example, *Physical Review* A 33, 759, January 1, 1986). No wonder that we think our mathematical form and arguments are so good. The early work was done by a Frenchman, Charles-Augustin de Coulomb (1736–1810), for whom the law of electrostatic forces is named. But the eventual development and culmination of an immense masterful elegant theory belongs to an unlikely British pairing, the English Michael Faraday (1791–1867) and the Scottish James Clerk Maxwell (1831–79), a common plain self-educated son of a blacksmith with no mathematical background or ability (Faraday) and a nobleman (the Scottish variety, a laird) who excelled in mathematics at Cambridge (Maxwell). They were both very dedicated, serious Christians, but even there the differences are striking.

Michael Faraday was the third of four children, born in a London slum to a family who had recently moved there from the rural countryside in the north of England due to economic pressures. He stayed in school barely long enough to get the ABCs of RRR (readin', (w)ritin', and (a)rithmetic). We would probably now call him a victim of bullying, a poor country kid who was made fun of in the city schools. In any case, his mother mercifully took him out of school. At their level of poverty, any apprenticeship leading to employment would be considered a blessing. But his rich religious life (the family were Sandemanians, a sect somewhat akin to Methodists) and a sharp disciplined mind and lifestyle prepared him to catapult above the meager opportunity afforded a thirteen-year-old by a bookbinder's apprenticeship. Books were available and he made excellent use of them, learning to take meticulous notes. The book business and the newly formed City Philosophical Society gave him excellent contacts. The secular world might call it fate, but a Christian would have to consider it to be God's hand in circumstances and disciplined preparation. But, in any case, the next apprenticeship, at age twenty-one, was with Sir Humphry Davy at the Royal

Institution, as assistant to England's greatest chemist. This was perhaps the most obvious leap in the steps from the ashes of a poor smithy's home to becoming a much-loved, highly decorated prince of natural philosophy who was eventually offered almost all positions of honor which were available. In all his storied career, he remained centered in the spiritual life of the Sandemanian church and his humble beginnings, "plain Michael Faraday to the last."[1] Most wonderfully from my perspective, his natural philosophy (science, as we call it today) and his faith were seamlessly united. As his agnostic colleague Tyndall wrote, they "could not be kept apart; there was an habitual overflow of one into the other."[2] Faraday wrote, paraphrasing St. Paul from Romans 1:20, that "even in earthly matters I believe that the invisible things of Him from the creation of the world are clearly seen."[3]

His down-to-earth but quite spiritual approach, coupled with his limited mathematics, made him uniquely suited to tackle the world of electromagnetism. He strongly adhered to what could be seen in the laboratory, eschewing speculation, but developing a visualizable model for his own understanding. As Maxwell later observed, this approach was deeply mathematical. It intuitively reverted to the ancient concrete mathematics of geometry which had been developed more abstractly in Descartes's analytic geometry and Newton's calculus. Nineteenth-century development of vector calculus and its masterful application by Maxwell in treating Faraday's vision eventually led to Maxwell's equations being deemed a paraphrase of God's first creation words, "Let there be light." But that's jumping way ahead.

Faraday's visualization of what he called lines of force can be most readily visible by the alignment of iron filings around a magnet. Iron filings become little magnets (induced magnetic dipoles, we say) which align along Faraday's magnetic lines of force (we now call them magnetic field lines).

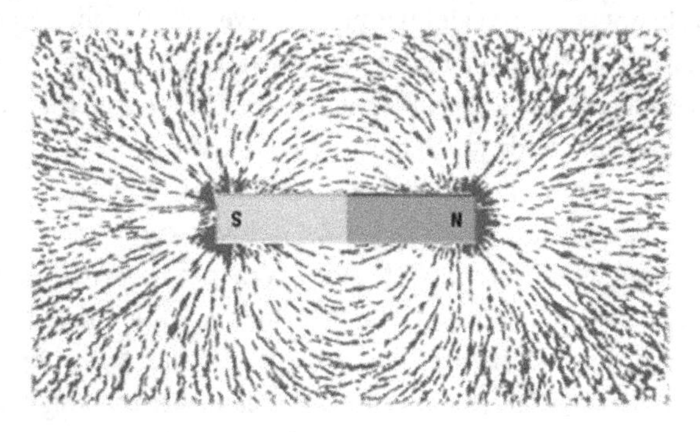

1. Zayonc, *Catching the Light*, 125.
2. Ibid., 126.
3. Ibid.

This is an example of the invisible being clearly seen, Faraday's Romans 1:20. Plus electric charges sent out electric field lines, minus electron charges

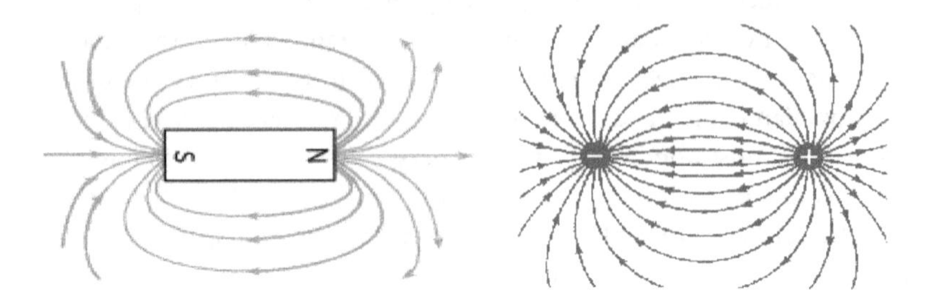

took them in (picture on right); magnetic north poles sent out magnetic lines of force and magnetic south poles took them in (picture on left). Thus the non-speculative, non-mathematical Faraday "saw" atoms not as hard impenetrable balls, the idea of mechanics, but as centers of lines of force; Nature was a sea of forces, intersecting, interacting rays which constitute a woven fabric of the whole universe. What started as a useful fiction, a mathematical crutch to a genius of limited mathematical training, became to him and later to the whole world of physics a reality which rivaled (or maybe replaced) the material image.

A key discovery for Faraday and for the world was what physicists call *electromagnetic induction.* Fascinated by the connections recently found between sound and light (by Young and Fresnel) and deeply convinced by his faith in the unity of Nature, he sought the connection in the concept of vibration in electrical effects. Thus in August of 1831, he discovered an electromagnetic action at a distance, transmission from one "vibration" through space to another, like the rattling of a violin hung on the wall when a nearby (but clearly not attached) violin is played. (We call this phenomenon *sympathetic vibration* or *resonance.*) He made two "electromagnetic instruments," which we call electrical circuits: one was a wire wrapped as a coil and attached to a battery with a switch to open or close the circuit (right side of the picture below); the other circuit had a similar coil, no switch, no battery, but a sensitive galvanometer (device for detecting current) completing its independent unattached circuit (left side). When the switch was opened or closed (a change which acted like the plucking of a string to cause vibrations) the galvanometer registered a surge of current. A change in one circuit induced a change in the other. We call it *electromagnetic* induction rather than *electric* induction because it was well known that the electric currents in a coil-shaped wire produced the same magnetic effects which we knew from natural magnetic materials called lodestones. It was the genius of Faraday to sort out and picture an interacting pairing of electric and magnetic lines of force. The *fact* was simply that flipping a switch in one circuit resulted in a needle jumping in a separate circuit. He supplied the

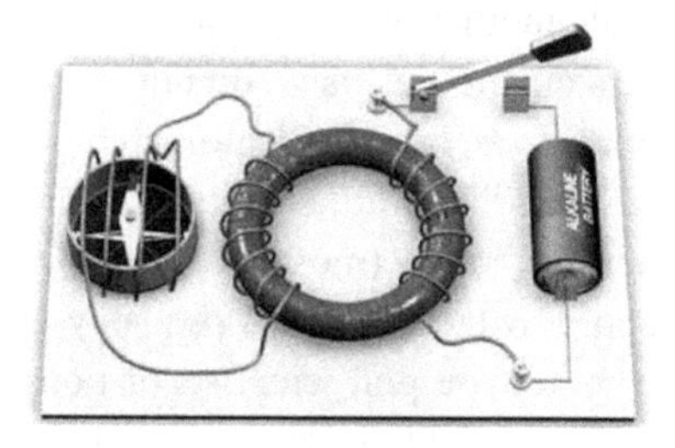

A simple laboratory experiment which crudely reproduces Faraday's experimental discovery of electromagnetic induction. The compass on the left acts as a crude galvanometer which can detect current. The iron core (the black doughnut in the middle) is not necessary, but increases the effect so that the crude galvanometer can see it. This you may obviously try at home. You may increase sensitivity by adding more coils around the compass or core.

invisible picture: changing electric lines of force changed current which changed magnetic lines of force which moved or had influence across space causing electric lines of force which caused current in the other circuit. A whole skein of *potential lies* to put together an indubitable *truth*: the needle moves. Every transformer on every telephone pole or electric train set, every radio or television transmitter and receiver, and now every cell phone screams out the fact which Galileo was supposed to have whispered under his breath after allegedly being forced to affirm the Earth's immobility at the center of the solar system: *Eppur si muove.* (Yet it moves.)

In Faraday's day, atoms were a time-honored convenient fiction dating back to the pre-Socratic atomists but were now finding some use by chemists to make sense of the exact proportions in chemical interactions. If two parts hydrogen and one part oxygen came from one substance called water, maybe each *piece* of water was *made of* two *atoms* of hydrogen and one *atom* of oxygen. We were not close to "seeing" an atom or in any way quantifying its size. Faraday dematerialized it further and suggested the invisible atom really consisted of some complex of invisible lines of force of two kinds, electric and magnetic, which could be converted into each other and propagated through space. That really seems like a hard sell or even a total con game. But Faraday's vision eventually made it more believable than Newton's totally unexplained and apparently unexplainable gravitational action at a distance.

How can you get across the chasm between God and man, however that chasm was produced? You need an intermediary, something or someone who belongs on both sides. That's what Faraday produced conceptually, a line however thin or imaginary which hung between two objects otherwise disconnected. Newton knew that no one could or should believe that one mass somehow reached across space to pull on another, millions of miles away or even mere feet away. We yearned for a connection and Faraday's imagination provided it. What he called lines of force, we call fields; they are the same thing. A mass (or a charge) is not only at some point in space; it carries or produces a field which extends throughout all space, all of the universe. We no longer have to say a mass reaches out and pulls another mass with no means; a mass produces an extended field which acts on another mass, on any and all other masses. This brings comfort to our minds, however legitimately you may think that we have been conned. It *seems* like a little trick; it *is* a little trick, but it is one that is firmly believed by many (most? nearly all?) physicists. Electric fields, magnetic fields, and gravitational fields probably all have higher rates of belief than God, whether you poll scientists or non-scientists. We do not see electric fields, magnetic fields, or gravitational fields, but we see events which we attribute to them. We do not see God but we see events which we attribute to God. The invisible is "seen" from the visible, whether it is in science or in religion.

Maxwell (1831–79) was a brilliant counterpoint to Faraday. He was the son of a distinguished family in Edinburgh, Scotland. At an age (nineteen) when Faraday was just getting introduced to science, Maxwell was already experienced with not only the experimental world of physics but had also published three mathematical papers. He had spent three years at the prestigious University of Edinburgh, reading widely in literature and philosophy, writing poetry, bright enough to do well in school and still be an untamed, unpolished free spirit. Sent to be tamed by the "Society and Drill of Cambridge," the big leagues for mathematics in England, he came in second in the Mathematical Tripos, England's equivalent of the mathematics Olympics, and a joint winner in the Smith's Prize, a competition among the best students in the Tripos, like an Olympics for Olympic medal-winners. At Cambridge, he was invited to join an elite discussion group called the Apostles, a society which over the years included Alfred Lord Tennyson, Bertrand Russell, Ludwig Wittgenstein, and John Maynard Keynes, to give a short list of some you might know.[4] The contrast of Maxwell's privileged education with Faraday's humble start could hardly be starker. Maxwell's faith was as clear as Faraday's but was expressed in private poems rather than the straightforward open prose of Faraday's lay sermons; both lived their faith by their actions. Although the rarefied air of Cambridge was officially quite religious (requiring ordination in the Church of England for most long-term appointments), Maxwell from his youth felt a schism between God and Nature and never became comfortable to publicly declare a narrowing of the gap. He felt that any attempts at harmonization of the "universal and public Manuscript" opened by God's servant Nature and "the written one of God" (the Bible) were to be personal.[5] It was the safe and convenient

4. Forbes and Mahon, *Faraday, Maxwell, and the Electromagnetic Field*, 145.

5. Zajonc, *Catching the Light*, 144.

position taken even by the great theologians, Luther, Calvin, and Barth, to name a few. This allowed Maxwell to be a sincere elder in the (Scottish Presbyterian) church but also to mathematize physics which makes God look like the great mathematician, a great and perfect law-giver who can lose his personhood in the Faustian bargain. (This is why Einstein can be the great admirer and follower of Maxwell and still not follow the personal faith in a personal God.) God took that chance when he gave Nature its freedom. (Please note that mathematicians can be and are persons, but when they get deeply into mathematics, you can certainly wonder if they are. Maxwell's Aunt Jane would try to jolt him out of such states by saying, "Jamesie, you're in your 'props' [mathematical propositions].")

Maxwell worked from the scaffold of mechanics of a material medium and ended by building a marvelous edifice of the immaterial lines of force of Faraday's imagination. Charges produced electric fields, moving charges (and magnets) produced magnetic fields. Faraday's law of induction meant that changing magnetic fields produced electric fields. All this went into a set of equations, the mathematical laws of the intermediary concept of fields which ruled the interaction of charges, which also always had mass, keeping them tied to a material world. Here are the equations in words:

1. Electric field lines are produced by charges, starting on positive charges and stopping on negative charges.
2. There are no known starting and stopping points for magnetic field lines. What we call magnetic north (N) and south (S) poles cannot be separated. (In our previous picture with magnetic field lines starting at the north pole and ending at the south pole, they in fact go right through the magnet, completing a loop.) Breaking a magnet in two produces two magnets, each with a north and south pole.
3. Currents (moving charges) produce closed loops of magnetic field lines.
4. Changing or moving magnetic field lines produce closed loops of electric field lines.

But Maxwell's coup was to produce mathematico-physical consistency and symmetry, by "seeing" that changing electric fields also produced magnetic fields. Even though they were in the immaterial world of fields where there may be no matter, changing electric fields acted like currents, he surmised, calling them *displacement currents*. This was a matter of faith and symmetry, unverified by experiments for several decades.

Thus the inventions of Faraday's fertile imagination concocted as helpful fictional children of the mechanical world have the very real potential, at least mathematically, to leave their parents of the material world and fly on their own. That is, we can leave charges and their movement (currents) out of the equations above and still get a self-sufficient pair of entities, called electric and magnetic fields. Taking down the mechanical scaffolding (which took Maxwell a long time and others even longer) does not cause the new immaterial building to collapse. Charges (stationary and/or moving) may tell of origins but it appears that we can leave them behind, producing two remnant laws.

1. Changing or moving of magnetic field lines produces closed loops of electric field lines.
2. Changing or moving of electric field lines produces closed loops of magnetic field lines.

Bye-bye, charges. Thanks for the creation, but we are flying the coop. Now you can see that electric and magnetic field lines can act as the chicken and the egg problem. They act as intermediaries for each other. Eggs are the chicken's way of producing other chickens, and chickens are a method for eggs to reproduce. We can think of generation skipping and write a single equation for how electric fields reproduce themselves in space and time and the corresponding equation for magnetic field lines. And then you will see: Voilà, two miracles occur. The equations for electric field and magnetic field propagation are exactly the same and the *one* equation implies they travel (together) at the speed of light, a well-known number but with no previously known relationship to electric and magnetic fields. (We might also consider as a third miracle that the electric and magnetic fields were transverse, perpendicular to the direction of travel and perpendicular to each other, giving us a model of polarization.) Thus electricity and magnetism and optics are suddenly unified in one theory. This is magic of the first order. Einstein considered Maxwell's work the greatest alteration in the conception in the structure of reality since Newton's foundational work. It is breathtaking in its scope and power. The first big expensive physics book I owned was informally called the bible of optics[6] and was summarized on the first page by all that was needed for the following 800 pages, Maxwell's four equations.

Maxwell and Faraday, in spite of their common faith, had quite different conceptions of the world. Faraday, unable to do the mathematical imaging, had "materialized" the fields, making them the primary reality. Everyone else, including Maxwell, clung to a conception of Nature as made

6. Born and Wolf, *Principles of Optics.*

of eternal bits of matter inhabiting space, historically and reasonably called atoms, indivisible fundamental building blocks of matter. The atoms were as unknown, unseen, and as speculative as fields or lines of force, but they carried the Newtonian agenda of a world made of moving, variously combined and related pieces of matter. They of course also harked back to the ancient pre-Socratic atomists. Faraday had neither the conceptual or mathematical power to construct a model of masses made up of lines of force, but that was his vision, a world made up of *real* lines of force or fields, wadded or knotted into little pieces called atoms and also filling all space.

Maxwell had great powers of mathematical modeling and a clear understanding and statement that all of his "theories" were *analogical* models. He believed that atoms were hard, impenetrable, eternal, unchangeable and impossible to wear out. They were the pieces God created first or they were givens for God. And they served as Maxwell's mechanical framework for the mathematics of fields which Faraday visualized. His structure was an incredibly complex mechanical model with vortices, cells, incompressible fluids, contraction, stresses, strains, inertia, wheels, gears, elasticity, essentially anything and everything that Newton's clockwork image had evoked. Maxwell was capable enough with the mathematics that he could conceive of it as a *model*, without investing it with *reality*. We say it was a mechanical scaffolding for building a theory of fields, but it was so huge, cumbersome, and so endowed with the spirit of Newtonian mechanics that the *real* building which Faraday envisioned was hid from view for several decades. The weird, awkward, and complicated framework of Maxwell could not be quickly and easily accepted. When in 1862, his theory, with the numbers of electromagnetic experiments, gave a velocity of electromagnetic propagation almost (within 1 percent) exactly that of light, the ungainly and bizarre construction hid the revolutionary achievement even from Maxwell himself. It could not be *reality*; it must be some coincidence.

Maxwell had displayed genius in the ingenious, albeit ungainly construction of what appeared as a mechanical monstrosity, much too complicated to be taken seriously. But more importantly, his genius was to truly follow Newton in his *belief* that the action at a distance needed explanation. Newton knew that the instantaneous action of one mass on a distant mass (and the later equivalent of a charge acting on a distant charge) deserved, yea required, some explanation. He considered such an idea, that an object can act on another through space (vacuum) without an intermediary or cause, "so great an absurdity . . . that no man . . . can ever fall into it."[7] Yet

7. Letter of I. Newton to R. Bentley, 25 Feb 1693. Quoted in Jeans, *Physics and Philosophy*, 116.

he knew he could not explain it; so he punted, *non fingo hypothese*. Maxwell was capable of constructing a bridge across the action-at-a-distance fallacy, even if it seemed to be a monstrous construction in its mathematical model. Faraday had by this time fallen into senility and probably was incapable (even in his best mind) of reading Maxwell's papers. He probably would not have approved. He believed in real, fundamental, free-standing, self-sufficient lines of force and Maxwell's machinery had demoted them, occluded them. Faraday's new vision pulled away from the mechanical model. Newton had knowingly recognized the gap and papered over it by a thin mathematical relationship, a simple mathematical formula, a numerical answer with no explanation. Maxwell, with the help of Faraday's vision and the electromagnetic relationships, had built a monstrous scaffold that loomed over the simple relationships.

When one takes seriously *both* the material world of fundamental particles *and* the seemingly ethereal world of Faraday's vision of lines of force, we have essentially kicked out the *analogous* mechanical model of Maxwell's theories without getting rid of the *real* mechanical world of Newtonian particles. It is then that Maxwell's equations, shorn of the complicated mechanical modeling, appear as a shining simple architectural marvel of Faraday's vision. Thus was produced a new world of fields, an additional reality for classical physics which mediated the interaction between particles, making the vague, even unacceptable forces acting across space, into an alternate, complementary world completing the unfinished business of classical mechanics. It was this that Einstein saw as the greatest alteration in the conception of the structure of reality since Newton. It is an obvious precursor to Einstein's further substitution of curved space-time for Newton's flimsy but world-shattering mathematical formula for gravity's action, properly hailed as "the universal law of gravitation."

We have often accepted the mathematical representation as something of a higher reality, without concern for the reality of the models or analogies. The concern for spiritual reality, held by those of us who see the emphasis on a mechanical world as at least a potential exclusion of spirit, can take a number of forms: 1) Berkeley's belief that the mind (ultimately the mind of God) is the reality which creates the physical; 2) the Platonic idea that Form or ideal or mathematical truth is the reality, the good, the truth, while the material world is bad, false, our bondage; 3) some duality game in which spiritual and physical live side by side with some ill-defined connection or perhaps a lack of connection. In any case, the reality of fields, in some kind of ghostly analog with spirit, seems to hold out hope for a spiritual reality or connection that was so missing in a mechanical model of moving masses.

Taking a different tack, the analogy can also be seen as providing analogies for the Old Covenant and the New Covenant, for an Old God and a New God. Action at a distance along with a complete law and order of mechanics provide a model of the Old Testament law. Faraday's deep belief in the reality of the lines of force gives a greater emphasis on the spiritual. A field as an intermediary is like an incarnation which connects to both ends, the heavenly becoming flesh and blood. God *could* act at a distance, but for some reason, he chose or it was in his character to provide an intermediary, the Messiah Jesus. We can believe God acts across space and time, across a chasm from heaven to earth, even believing that the Law is a means of that grace. But we want, perhaps we need, a further explanation or connection, in the personal Mediator, Christ Jesus, and in the further mediation of the Holy Spirit when he, Jesus, is no longer here in person.

Theologians, irked by the power, universality, precision, and total dominance of a mechanical world of matter in motion, Newton's seventeenth-century invention, which seemed to leave no room for spirit or God, can obviously be sucked into the immaterial world of fields, a seemingly clear link to an omnipresent spiritual world, a world more palatable to a religious mind. But beware the old and new temptations. The problem most of us have is the duality of immaterial fields alongside the material substance, of the spiritual alongside the physical, decoupled pairs in both cases. The two have to be linked, and the fields and the spiritual realm have a way of flying off on their own.[8] How, why, and to what extent God decoupled himself from his creation may remain eternally unanswered or unanswerable questions for us. But the Christian message is and must always be that God did not completely decouple himself; he is not separate and he will not be separate. The Deist God, a God who created and *walked away*, is not the God of Christianity. In my opinion, it is not the god of any religion worth living and dying for. I will say that it can be the god of wonder and even worship for a scientist like Einstein and, yes, even Dawkins, but it is not even close to being the *whole*, not to mention *holy* God.

The possible decoupling of light from matter as we see in Maxwell's equations has very real consequences in our world. It means that light can move in what we call a vacuum, a place of no matter. This means that the velocity of light, that mysterious very large number that pops out of our electromagnetic work is an absolute, independent of the mechanical platform we live on to make any such measurements. We call such platforms "reference frames" and light travels in a vacuum at $c = 3 \times 10^8$ m/s in *all*

8. St. Augustine engineered the long-standing theological divorce between spirit and matter by declaring that what does not do anything to save the soul does not matter.

reference frames. That is the basis of Einstein's theory of relativity which makes meter sticks and clocks *depend* on the reference frame, be *relative* to reference frames so that the absolute c holds. This is again a wonderful truth about God as the Absolute; Einstein's theory should be called the theory of absolutes, or *Absolutenprinzip* as he called it in German. But we will always find it difficult to see *what* in God's creation reflects that concept and *how*. There is plenty of room for mistakes. Almost all of us think that measurements of space (meter sticks) and time (clocks), or equivalently space and time themselves, should have that honor of absoluteness. And there are plenty of other choices we can try.

With the completion of Maxwell's epic work, fields and waves (vibrations which travel) in those fields took their place as the second major pillar of what we call classical physics, a traditional second (semester) course in teaching you all you may not want to know about the physical world. Let's review. Mechanics told us we could figure out everything about the motion of all the pieces of the universe if we just know the mathematical formula, not the *how* or *why*, of force laws. Field theory added an immaterial model for the forces, almost like a parallel universe. Newton's substitution of a mathematical form for an otherwise mysterious force acting at a distance had (with Maxwell and the electromagnetic forces which mimicked the gravitational form) morphed into a theory of intermediaries which we called fields. Two problems arise. One is the decoupling which is possible. Secondly, at least a problem from my perspective is that *both* worlds are deterministic; the machine and the field-ghost in the machine are still just a machine. It, the combined machine, just follows the rules, perfect laws laid down by the Creator or by some principles above the Creator. You may think that you could have made the world better than God (or at least told him how to do it better), and having everything obey the rules is likely part of your plan. If you want to make *machines*, you're probably right. If you want to make something that's in your image, at least alive, you may well be wrong. By putting fields in, we've added fuzziness to the sharp picture of classical mechanics, but fuzziness in the head is not a mind and fuzziness in the liver is not a soul and a fuzzy aura around your body is not a spirit.

The third great leg of classical physics is thermodynamics and we will tackle that next and see if we can start putting these disparate approaches for disparate entities back into one world.

6

Turning the Heat on a God Who Is Too Big

Looking at the *internal* unseen world of heat yielded thermodynamics, and Planck's thermodynamic exploration of the continuous electrodynamic world produced evidence of discontinuity, an atomicity of light.

WHEN THE WORLD IS apparently made of pieces too small to see, there is plenty of room to hide mysteries. Heat was one of those mysteries which we could not see, but we could feel. Objects could be heated or cooled over a fairly wide range of temperatures without appearing to make any changes in mass, color, size, or shape. Like the ether, the substance involved in heat flow or causing heat sensations must be some special type of medium, phlogiston or caloric, as it was variously called. It can apparently come from hot objects and go to cooler objects as though it is a substance which flows, but it also appears to be manufactured by rubbing. Count Rumford (see Benjamin Thompson–1814) was called Benjamin Thompson (1753–see Count Rumford) (both pictured here) when he married the reverend's daughter, recently enriched by widowhood, had a daughter,

and then deserted them and America during the Revolutionary War. (Is this the first American scientist we've mentioned? And *he* betrayed us!) Using his scientific expertise and personal charm to curry favor by aiding the military efforts of European nobility, he achieved his title at court; he earned his place in science history by noting that heat was none other than motion. This was not a casual half-hearted throw-away comment. Whether it was spying, military exploits, or currying favor with the powerful, the rich, the titled, or the many women, especially the powerful or rich or titled or widowed women, he did not do anything half-way; after all there are at least two sides on which to work if you don't have strong loyalties. One of his greater contributions to science was his establishment of the Royal Institution in London in 1800. Setting Humphry Davy at its head, he thus indirectly provided the young Michael Faraday with the opportunity which catapulted him into the ranks of science.

Heat was a major theme in Rumford's career, ranging from the mundane but very important kitchen range and soldiers' uniforms (by no means unimportant) to explosives and the manufacture of firearms. Early in his career, he had noted that a cannon fired without a cannonball had been heated more than a loaded one. This piqued his interest, making him suspicious of the explanation of heat as a fluid, whatever it was called. When he became involved in the boring of cannons (pictured below), he made a long and involved investigation of the heat process, literally boiling water by carefully measured horsepower (real horses) and trying to weigh the supposed caloric. He came to the conclusion that heat could be nothing but motion, albeit an invisible motion hidden inside the matter. Thus began the further take-over of physics by *matter in motion,* a mechanical view of the universe which certainly seemed destined to win.

Because of its hiddenness, tucked into the internal workings, heat was destined to spawn a new discipline, a third rail of classical physics. This discipline explicitly limited itself to *systems,* looking at external parameters to deal with something that might be going on inside beyond our sight. *State variables* such as pressure, volume, and mass could be used to specify the *state* of the system. Mechanics had dealt with extended collections of matter but treated their motion as a whole like the movement of a point (the center of mass) and rotation of a rigid body, or as relative motions like compression, extension, twists, or vibrations that could be seen. Thermodynamics admitted an unseen internal being while trying to make laws about an entire external system.

The first hurdle was to formalize a meaningful quantitative measure of temperature, what we could feel as hot and cold. The direction (hot as bigger numbers or the reverse?), the magnitude (how big is what we call a degree?), and the starting point (how do we define zero degrees?) of a scale were all done arbitrarily with no regard for or knowledge of fundamental aspects or laws of nature. Thus we have two scales (Fahrenheit and Celsius) which go in the same direction (hotter is bigger numbers) but differ in the zero and in the size of a degree. The story is that 0^0F was chosen to be the temperature of a particular winter day in Gdansk and 100^0F was his (Fahrenheit's) body temperature. With such possibilities, it is no wonder that there are several other scales which have only historical interest. For scientific purposes, there are only two ideas needed: a reproducible and definable measuring process and a meaningful fundamental idea for defining a zero-point. The measurement depends on a simple idea of *thermal equilibrium*. If two systems are allowed to interact thermally (i.e., heat can flow between them), they will come to thermal equilibrium, which means, by definition, that they have the same temperature. *If system A and system B are in thermal equilibrium with system C, then they are in thermal equilibrium with each other.* This seems so self-evident that we neglected to make it explicit, until other laws had already been formulated. We already had a First Law, so this became the *zeroth (0th) law of thermodynamics*. Thus any measurable property of a system designated as system C which changes with temperature (e.g., height of mercury in a mercury-glass thermometer) can act as a thermometer, allowing us to know when $T_A = T_B$ by checking that $T_C = T_A$ and $T_C = T_B$, even if system A and system B cannot be brought together (because of space or time).

The first law of thermodynamics is essentially a law of conservation of energy: *Energy is neither created nor destroyed; the total quantity of energy remains constant.* Kinetic energy (given by (1/2)mv^2 where m is mass and v is velocity) and work (force times distance) had by now prominent positions in the classical world of mechanics. They had the same units or dimensions ($[ML^2/T^2] = [ML/T^2][L]$) and we say that work can change the kinetic energy. A force pushing through a distance can speed up a mass. But Newton's third law of motion says that the mass pushes back equally through the same distance and thus does an equal negative work on the pusher. Thus there is *no total net* change of energy. Thus classical mechanics developed a faith in total energy conservation in the universe or in an isolated system. The game was simply to make sure we found all forms of energy (essentially anything which could be measured or produced by [ML^2/T^2]. When we did find all the forms (essentially the same game as finding all the forces), we were rewarded with a total sum of energy which did not change in all the processes. If it fails us because of hidden motion or forms, our faith leads us to look harder or even make up what we imagine might be some hidden form. Even if we don't know what it is or how it works, it may still be a useful concept. Thermodynamics dealt with systems by their exteriors, openly admitting that there was an interior which had hidden features. Thus both heat flow and internal energy became useful concepts: heat flow was an energy transfer driven by temperature differences; internal energy was a hidden energy content which was presumed (by Rumford's suggestion) to be motion of the inside parts which could not be explicitly seen. Energy conservation was a good useful principle of the mechanical world as long as we admitted there were *seen and unseen* energy transfers (those due to the forces we could see and follow and those due to unseen forces) and *seen and unseen* energy contents (the classical mechanics ones of the exterior system were seen and internal ones were unseen). I know you would not be happy if your bank worked this way, with one set of books for the auditors and another one hidden inside, but thermodynamics did very well, finding the conversion factors between the ideas of work measured in standard units (called *Joules*) made from standard measures of length, mass, and time and a subterranean world of thermal units (calories, degrees, etc.). The concept of energy conservation, in spite of its cobbled construction through the centuries, has become ever more important in the world of physics. My physicist friend has said that he would prefer to lose his right arm than to lose the conservation of energy. This may rightly reflect a principle of God's faithfulness and constancy. Emmy Noether proved that in our physics/mathematics system, every symmetry of our universe has a conservation law associated with it. Conservation of energy is thus bound to what we call *translational symmetry in time.* It basically says that all physical processes, any experiment you want to try, comes out the same no matter *when* it is done. This is "Jesus Christ, the same yesterday, today, and forever" extended to all of creation.[1] A great general idea but be forewarned: not every supposedly obvious symmetry holds. We can be fooled. But symmetries and their related conservation laws have continually grown more important while becoming more abstract, complex, and mathematical. And we can also be

1. Hebrews 13:8 KJV.

reminded that the ultimate symmetry is an undifferentiated nothingness, not so obviously interesting.

Energy also provides a natural idea for a zero of the temperature scale. When low-density gases are cooled, we find that they have pressure *proportional* to temperature (for all different kinds of gases) *if* we use -273.15°C or -459.67°F as a zero of a new scale called Kelvin or absolute temperature scale. Since the pressure of gases is presumed to be due to the collisions of the moving molecules with the walls, the pressure goes to zero when the molecules stop moving. Thus a fundamental idea of minimum energy occurs under special circumstances which we call *zero Kelvin*, 0 K, the motionless world.

The concept of a world with a total constant energy should perhaps suggest that we could trade energy around in its various forms at our convenience. Why not take the hidden internal energy from the dirt outside our house to give us warmth inside? After all, it is a very long way from 0 K. Indeed, why not take heat from anything to convert it to kinetic energy, the motion we desire for getting around? Making something warmer (our heaters), making something colder (our refrigerators and air-conditioners), and making something move (our vehicles) are just matters of moving energy around and thus do not change the world and should not really cost us or the universe anything. You might recognize that the world does not work that way, and any decent science of the thermal world has to understand *how* or *why* it has some limitations. And classical thermodynamics stated a basic law that does exactly that in terms of state variables, the big external parameters of the system, without knowing the invisible internal world. To do so, it defined a quantity called *entropy*, the Greek root word *tropos* meaning *direction*. This quantity basically tells the *direction* Nature takes and the limitations we have in taking any direction we might like. Changes in entropy involve heat flows and temperatures. The simplest form of this law is: heat always flows naturally from the hotter to the colder. To state it more carefully: a closed system involving only two subsystems (A and B) at T_A and T_B respectively and no physical processes other than heat flow will only have heat flow *from* A *to* B if $T_A > T_B$. We can concoct a complicated set of machinery with various barriers and valves between them; no matter how sophisticated we get, our laws must tell us what is possible and what is not. *The second law of thermodynamics* can be simply stated: *In a closed system (no external influences), the total entropy of the system increases.* (Whether we define entropy to increase or decrease is a choice, just as temperature scales have only the defined direction we put into them; we have chosen the definition so that the direction of this quantity is always up.) This law limits the operation of our heating and cooling systems and the efficiency of our heat engines. Just as c is the absolutely enforced speed limit of the universe, the second law of thermodynamics is the enforced banking system of energy exchanges. Taking heat from anything in the universe has its costs, which will be paid, willy-nilly, i.e., whether you want to or not.

Whether we like it or not, the world's energy accounts are graded basically by the temperature scale and temperature differences are what drive the world. The direction of the world's process is toward thermal equilibrium, the hot cooling off and the cold being warmed, until the ultimate endpoint: everything is at the same temperature with no more driving potential. This is what we call the *heat death* of the universe. And it whacks us in the face with the elephant that has always been in the room of classical physics: it *never* included life *ever* anyway and now it threatens the end of *all* life, with no reprieves possible.

Entropy is often said to be a measure of order, and it *is* in lots of subtle ways. The crudest is that any two sub-systems with two different temperatures is ordered; when they are allowed to come to equilibrium, that ordering, a hotter and a colder, is destroyed and the entropy which we can calculate will be higher. Entropy increase is thus associated with loss of order.[2] A crude and dirty statement of the second law is that things always tend to disorder; your neat stacks in your office will get scattered. In this vein, it is often stated by creation scientists that evolution is disallowed by the second law of thermodynamics, which of course means that one is choosing between scientific theories, with the second law trumping evolution in this case. This is a total misapplication of the second law. If the second law says that evolution cannot occur because it increases order (decreasing entropy), then the second law also says that life cannot be, and we who are alive cannot order those stacks in our office before they ultimately get scattered. Entropy can and does decrease every day in multiple ways in multiple places. The second

2. Physicists think of the laws of thermodynamics as very general principles which govern *everything*. Thus they like to half-jokingly apply them to other areas of life. For example, the applications to economics with a temperature scale providing an analog for *richness* and heat flow providing an analog for capital flow might yield a "second law of economics" that capital always flows from the rich to the poor. Or it might yield the reverse: the rich only get richer and the poor only get poorer. Trying to stick to the ideas of our real thermodynamics, but express it in other language, I like to think in terms of competition. 0th law: "Let's keep score!" which means I think I play better than you. (Thermodynamics does this by defining the ordering parameter, temperature.) 1st law: "Your loss is my gain." It's a zero-sum game; I get what you lose. OK, it might happen the other way sometimes. (The real first law says this as a conservation law for energy.) 2nd law: "The end result is always a tie." For a competitor, this is like kissing your sister. And you simply cannot get out of the game. (In real thermodynamics, this is the inevitable direction and ultimate result of thermal equilibrium at the heat death of the universe.)

law only says that it must be paid for somewhere else in the universe (by corresponding entropy increases). Life does not exist by itself; it exists at the expense of something else. Our sun (a hot side) and space (a cold side) represent a heat engine which allows us life, our meager potential for ordering. We thus call the sun our primary source of *negentropy* (negative entropy changes). The second law says that life cannot exist *by itself*, as an isolated system. Try living in a sealed container and you will find out.

The power which thermodynamics holds is not just a crude one which shows the direction of an isolated universe, a universe without the input of Something or Someone. At least its supposed power within our experience comes from the mathematization, as so much of the power of physics does. Thermodynamics declares that these things called energy and entropy are what we call *state functions*, mathematical functions that can be calculated perfectly from the exact values of the *state variables*, measurable quantities, independently of how or when it got to that state. In other words, it carries on the tradition of Newton's calculus which makes claims of certainty, now in complex systems with lots of things hidden inside. Atomism may thus be a moot point. If we can carry on the calculation perfectly without knowing details inside, the inside may be as irrelevant as the absentee God, just unnecessary metaphysics all, and with no malice toward God or atoms.

Enter stage left, Max Planck, an upright man. What a wonderful testimony to be labelled "*. . . an upright man*" in the title of one's biography. But it wasn't easy, as the preceding words of the title "*The dilemmas of . . .*" will testify.[3] Born in 1858 in a privileged educated German family and living past the end of the Second World War, he saw and contributed to the glory of a rising Germany and, by his immense integrity and the respect of the international community, also contributed greatly to the rise of German science from the ashes after its annihilation by the anti-semitic Nazism and the war it generated. The old Kaiser Wilhelm Society for the Advancement of Science (*Kaiser-Wilhelm-Gesellschaft zur Förderung der Wissenschaften*) was reconstituted as the Max Planck Society in his honor and probably only because of the international faith in this very old man who was now reduced to

3. Heilbron, *Dilemmas of an Upright Man.*

running to the countryside to survive the war. He lost a son in the First World War, and quickly lost both twin daughters in childbirth (married to the same man, in succession) and finally lost his favorite and only remaining son by execution for his involvement in plots to assassinate Hitler. This is one good candidate for a modern Job.

Planck was born into a family of academics, theologians, lawyers, and churchmen. He was talented enough to enter any profession he chose, including music. Even though he was discouraged from entering the world of physics, he chose it because he thought it dealt with the deep universal questions. He chose a science based on *principles.* Although he was an exemplary student and worker with a good mind, he had good reason to think that his family connections and contacts played a big part in his academic appointments, and he was grateful. In both his choices and his work ethic, he acted as a public servant, seeking knowledge, disseminating knowledge, and using knowledge with a belief system strongly influenced by the deep Christian roots of his family. Even though the God he professed tended to the impersonality of Einstein's God, he understood and expressed openly and often that *faith* was an important part of his science.[4] He was uncommonly aware and reflective of the philosophical backings, right or wrong, of what was going on in science. While Einstein drifted, with great intuition but little reflection, through both positive and negative relationships with Mach's anti-metaphysical positivism,[5] Planck took strong and steadfast positions based on his experience and deep beliefs. When he grew old, he became an evangelist for his views on science and a revered grandfather figure who tried to hold Germany's great scientific tradition together while its economy, political system, and moral bearings fell apart. Although he never preached the gospel of a personal God, he realized in his final years the strength that his deep personal faith had given him. Facing the dilemmas that Hitler's Nazism presented to this beloved international man of science forced him to use all the resources of religious faith, scientific faith, his widespread scientific community, and a developed work ethic, all of which he had carefully nurtured.

His first major choice within physics was thermodynamics, believing that it stepped back from particulars and dealt with some universal themes. He became an expert and proponent of this discipline which upheld the

4. Planck said that "over the entrance to the gates of the temple of science are written the words: Ye must have faith." Lindley, *End of Physics*, 6.

5. Einstein said that a scientist must appear as an unscrupulous opportunist in his philosophy, being a realist, an idealist, a positivist, even a Platonist or Pythagorean for various reasons. Pais saw him as a prisoner of an objective realism for the last thirty years of his life. Pais, *Subtle is the Lord*, 13.

great energy conservation law and gave a direction of time, albeit a direction toward the ultimate death. Although classical physics had been founded on causality, usually stated in terms of temporal direction (cause *precedes* effect), the mathematical result was an abstracted time, a simple coordinate which was really the same going left or right in spite of our tradition of reading and thinking left *to* right. Quite simply, Newton's laws of mechanics were independent of the direction (or mathematical sign ±) of time. We often boast in modern Western civilization that the ancients and the East were stuck in time which did not go anywhere because it was *cyclical*, but the scientific revolution (our vaunted Western product) has given us a purely mathematical time which did not go anywhere because it went equally backward and forward. It had tried to capture causal time but had ended up with only the skeleton because it had failed to capture *teleological* time, or *eschatological* time, to use the more theological idea. It did not even capture the *personal* time of human consciousness, a directional time with past, present, and future.

Let's be honest and blunt; physics simply did not have *life* in it. It had left out Aristotle's *formal* cause, the Form or design which resides in the Creator's mind or in some Celestial Realm which may be the Creator's mind, the mind of God, and the *final* cause, the purpose or *telos*. Physics, when it finally blossomed in the seventeenth century, always avoided the organismic, holistic approach in favor of reductionism, the natural atomistic approach. Thermodynamics had recaptured (very weakly) the organismic by looking at systems whose details may not be known and got a historical one-directional time but a one-way road to death. Ironically, this occurred almost precisely at the time of Planck's birth and the publication of Darwin's *Origin of Species* (1858). The theory of evolution was the story of life writ large, a resounding affirmation that negentropy, order, is available and is being used, even if the price must be paid and the bank account is finite and *will* run out eventually. Thermodynamicists may have thought they could

calculate the entropy from the *physical* state, but it did not know how to account for the breath of life; a dead cow and a live cow had all the same physical state variables, therefore the same entropy, just before and after death. But *life* can appropriate negentropy from its surroundings and *death* can only decay. I am not aware of how Planck understood this, but he did see the broader systems approach of thermodynamics as an avenue to deal with more universal questions.

Most great discoveries in physics are the flaming red-hot leaps of young men, but Planck seasoned himself with his preparation in thermodynamics well past the over-the-hill thirty years of age for theoretical physics. While electromagnetic theory blossomed almost a decade after Maxwell's untimely death in 1879 from abdominal cancer and the world of radioactivity was just breaking loose, Planck was living in the stodgy world of thermodynamics, looking for an application which had some wholeness and universality. Heinrich Hertz (1857–94) had discovered radio waves in 1888, cementing the union of electricity, magnetism, and optics, confirming the continuous passage of energy in continuous space and time by continuous fields, a vision of Faraday's individual lines of force brought to continuity by Maxwell's mathematization and brought to the laboratory by Hertz and his fellow Maxwellians. A counterpoint to this new world of continuous fields which carry energy had been developing as well. Maxwell and Boltzmann (1844–1906) had dared to look inside the thermodynamic systems of simple gases, thus drawing the thermodynamic world back into a mechanical model of masses in motion. Atoms may have been a background conceptual model dating to pre-Socratic time and central to a Newtonian view, but the actual entities were unseen. The continuous space-time-causality coupled with the mathematics of the continuum had left matter treated as a continuum. Indivisible atoms (the original Greek word literally means *not* (Greek a-) *cuttable* (Greek -tomos) did not really figure in an infinitely divisible space-time mathematics. And thermodynamics caught the disease, developing with all the mathematics of calculus, *and* with all the certainty of classical mechanics. Maxwell and Faraday, eschewing the invisible action at a distance and fervently, religiously going to the laboratory for motivation and for confirmation and correction, had nevertheless produced an invisible world of fields which gained their reality by carrying energy and making sparks in Hertz's lab to confirm the ethereal radio waves. Maxwell and Boltzmann had begun to make material discontinuous atoms *real* by working out the velocity distributions of the molecules which made up a gas. By clever thermodynamic arguments, one could find the average kinetic energy of a molecule to be proportional to the absolute temperature T and could *calculate* the fraction of molecules at various velocities. As energies got larger (faster velocities), there was a factor, called the Boltzmann factor, which tells how the number of molecules decreases. Again, it was a nice mathematical theory and formula about things we could not see individually, building our ivory towers of theoretical physics. But Maxwell, remembered as the consummate theorist, was also a consummate experimentalist. Home, at his country estate Glenlair for the summer months, he performed the clever experiment with his wife. Single atoms or molecules could not be seen and counted, but *relative numbers* of them could be measured. This was a singular victory for two directions in physics: 1) the stranglehold of the mechanical view of Newtonian mechanics, and 2) a statistical approach which was creeping into thermodynamics. There was a simultaneous affirmation of the *certainty* based in Newton's laws and an admission that we could not know the details except by statistical arguments and measurements.

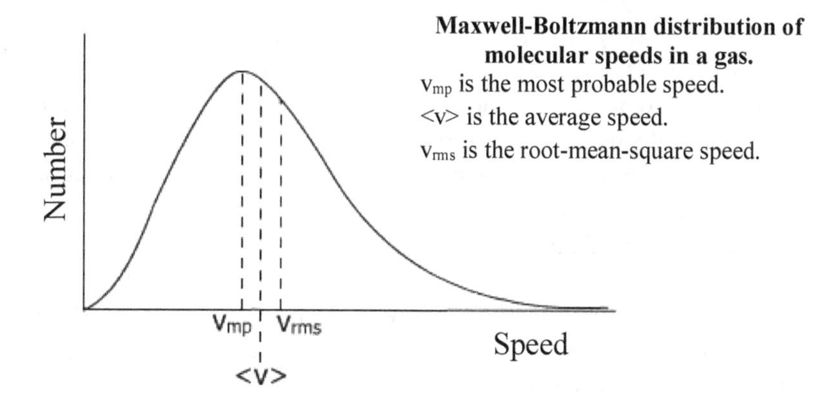

Maxwell-Boltzmann distribution of molecular speeds in a gas.
v_{mp} is the most probable speed.
$\langle v \rangle$ is the average speed.
v_{rms} is the root-mean-square speed.

Physics, particularly on the Continent, was in the grips of the philosophy of Ernst Mach (1838–1916), who took the critical and convenient view that physics was about what you could sense with your five senses. Faraday believed you could see into the *invisible* world by what you could see in the visible world. He thus put great emphasis on what you could see and test for yourself, but always with an eye to seeing beyond. Mach made science into an economic summary of what you could see and test for yourself. It's a good way to be sure you're right: Stick to the facts. But it takes the life out of physics. Anything beyond physics is metaphysics and is clearly off-limits for our good solid knowledge. In this view, religion is of course meaningless, but Mach's unfortunate debacle was that *atoms* also were meaningless. Don't believe in what you can't see. And he managed to live at precisely the time that atoms became meaningful. Although I am adamantly opposed to the positivistic philosophy of Mach, let's be clear that his advice is not all bad. The absolute space and time of Newton came under the same critique of metaphysical concepts in physics, and this motivated Einstein's theory of relativity. Einstein later said that Mach's philosophy was good for exterminating vermin, but could not *produce* real science, which required that creative spark, an imagination which saw the unseen, hauntingly close to a biblical definition of faith.[6] Einstein recognized that to be human was to be metaphysical. Mach rode his hobby horse too far and Einstein's contribution to atomism helped knock him off; well, it did not dissuade him but it certainly sealed the fate of his reputation.

Boltzmann felt as though he stood alone in the German-speaking world in his promotion of atoms, and Planck *did* stand against him. We can probably not say that Planck did not believe in atoms or that he was against them, but in his certain and continuous world of what we now call classical thermodynamics, you do not need to look inside and the discontinuity of the Boltzmann atomic world collided with a continuous world, conceptually if not practically. He certainly did not give comfort to Boltzmann's view, and Planck, ever the gentleman, found himself on the outs with Boltzmann, an unusual if not unique situation for Planck. Boltzmann, beset by bouts of depression and a tempestuous spirit at times, was not helped by his lonely position in the physics community and committed suicide at just about the time his views were being proven right.

Atoms, as we all know, have not gone away. But Planck had other fish to fry. The new world of electromagnetic waves, along with thermodynamics, should promise to explain the behavior of what was called a *blackbody*. All materials glowed red when heated and then turned white-hot at higher temperatures, pretty much independently of the material or the shape. An ideal material for this was black, absorbing all light that struck it, and an ideal black material could be made by a container with blackened walls and a small window. Any light going in would be absorbed by walls, or bounced off only to be absorbed by another wall. The distribution of frequencies in the electromagnetic radiation (we call it *spectrum*) coming from such a window was dependent on the temperature and nothing else, not the material, size, or shape. Here was a pure "gas" of continuous electromagnetic radiation which could presumably be treated by the continuous calculus of this quite general science of whole systems, thermodynamics. This had a universal ring to it, seemed to be decoupled from the material world, and he was an expert in the field of thermodynamics. Moreover, he had friends

6. An alternative statement of the relationship of Mach and Einstein with respect to relativity is: Mach said that Newtonian absolute space-time was meaningless, while Einstein said it was *wrong*. The positivism is safe but powerless. Einstein was willing to go out on a limb.

who made these kinds of measurements. He and the world were ripe for a revolutionary upset.

At first glance, the problem of Maxwell-Boltzmann's molecules bouncing around in a box with some coming out a window seems almost the same as Planck's problem of electromagnetic radiation bouncing around in a box with some coming out a window. And the results for frequency distribution in Planck's case look very much like the results of Maxwell-Boltzmann, at least crudely. But the difference in the "material" makes a huge difference in

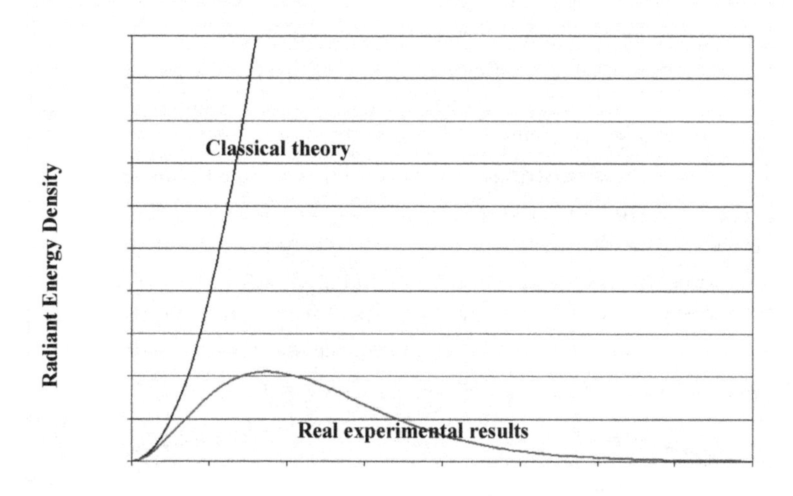

how you approach it theoretically: one is a finite collection of something like small balls, discrete entities, and the other is energy being carried continuously in a continuous space. The first has *individuals*, the energy of which is regulated by thermodynamics and has a specific average kinetic energy of (3/2)kT. (The 3 is because of three dimensions allowing velocity in each of the three directions.) k is called the Boltzmann constant and is clearly a conversion factor between the arbitrary temperature scale and the energy which it represents. Temperature T is a nice numerical rug under which we hide the energy of internal pieces of energy, whose nature may be unknown but whose averages we know. Here I must confess that the proof of this average energy depends on something we have been taking for granted, the *seemingly obvious* fact that the energy can take on *any* value; the molecule can move at *any* speed. The velocity "space" and energy "space" are as continuous as our real space and time. No set of two values, no matter how close, precludes a value in between. After all, isn't this obvious for everything? Well, two particles do not always have a particle between them, and that allows us to actually account for each and every particle.

In his case, Planck had a continuous space of continuous fields, as far as he knew by the theories of the day, and he needed a way to account for them in terms of frequency (or wavelength), because that's what we could measure: energy as it depends on frequency. Whether one counts frequencies or space, there is an infinity to account for; if each "piece" had energy (1/2)kT, then we have an infinite amount of energy. Besides, it cannot make sense that each "piece" of space has energy (1/2)kT; we could divide it into two (or more) and suggest that each of them should have (1/2)kT. Infinite divisibility has its problems. We need some way of having a countable number and having a cutoff for higher energies. The physics of music, going back to the pre-Socratic Pythagoras, gives us an idea of countability, getting back to discrete integers rather than the games of calculus.

A string (used in stringed instruments) with fixed ends was understood by Pythagoras to have discrete modes of vibration, called *standing waves*. It vibrates as a whole, in two parts, in three parts . . . as shown, producing what we call *harmonics*: f_1 (the fundamental), $f_2 = 2f_1$ (second harmonic), $f_3 = 3f_1$ (third harmonic) . . ., musically called *tonic, octave, octave + fifth, etc.* Now we have a way of counting, but it does not seem to have a stopping point. In fact it does have a stopping point if the string is made of

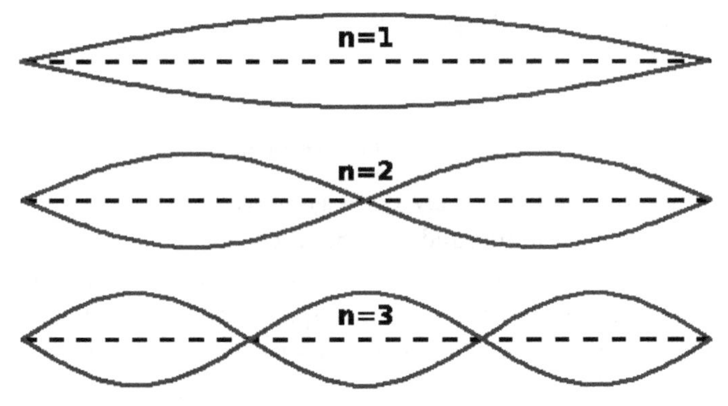

atoms equally spaced, because frequencies which have waves shorter than the distance between atoms are not new frequencies; they have already been counted. Three dimensions of a *volume cavity* (rather than the one dimension of a string) produce more complicated possibilities but the basic idea is the same. The unending infinite number of waves possible is terminated by the finite number of atoms.

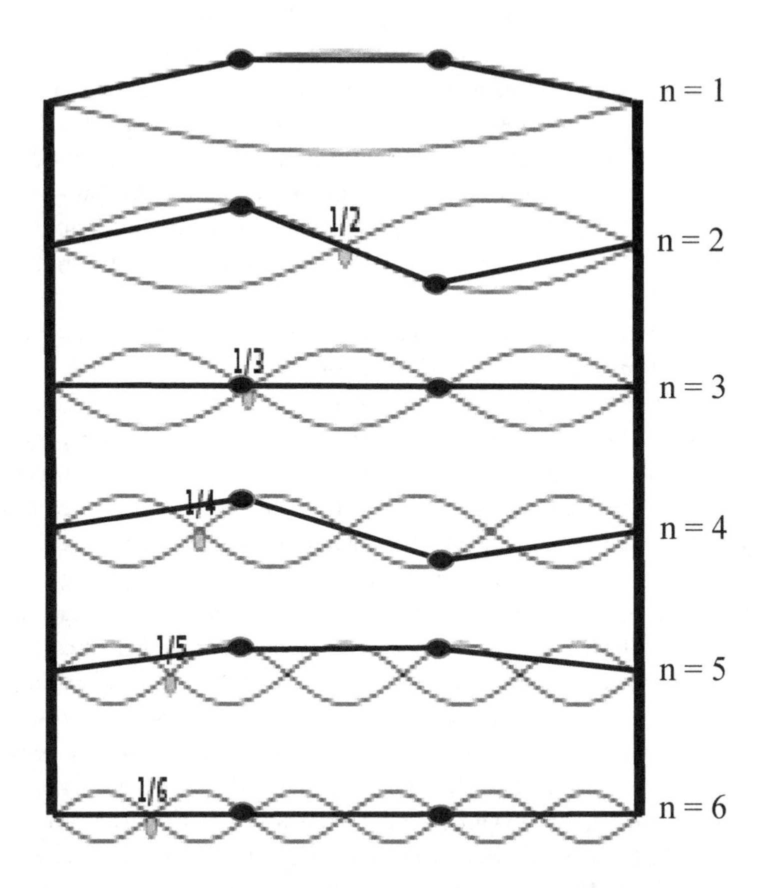

Two masses have only two modes: going together (n = 1) and going opposite (n = 2). n = 3 produces no motion. n = 4 reproduces n = 2, the masses going opposite. n = 5 reproduces n = 1. Shorter waves (higher n) just continue to reproduce the same two modes (or produce no motion, n = 6).

This is a great stratagem for waves in an *atomistic* material medium. *But* continuous Newtonian space and continuous Faraday-Maxwell fields have no stopping point *because* there is no stopping point in dividing pure space and time and thus no end to the counting process, no limit on the energy, no cutoff for the high frequency end of the spectrum (as there was a cutoff for high velocities). This was called the *ultraviolet catastrophe*, a theoretical fiasco because the naturally appearing theory suggested that higher frequencies (ultraviolet light is higher frequency than visible light) should have more energy, an infinite amount of energy as we went to infinite frequencies. (See the "Classical Theory" curve shown on the "Blackbody Radiation" graphs shown above.) Nature, the real thing, did not have infinite energy in a hot cavity, but the continued belief in continuity, continuous space-time, fields, frequencies, and energies had conspired to produce a theoretical impasse.

A typical calculus ploy is to divide your world of interest into a finite number of small intervals and then consider the possibility that the size of the intervals gets smaller as the number of intervals increases. This process can, in the mind of the mathematician, continue *ad infinitum*, and the mathematical process gets your desired result, perfectly accurately. Planck was a believer and tried this process with the energy possible in each frequency mode. Suppose, he says, the energies possible at a given frequency, f, are 0, hf, 2hf, 3hf . . . , forming an energy ladder of rung spacing hf with an infinite but countable number of rungs. Now we let the rung spacing get smaller; i.e., let h get smaller. When h gets to zero, our mathematics should give us the perfect right result, the way Nature behaves, *if* everything we have put into our calculation is right. Nature had provided the answer in the back of the book, and Planck's friends had carefully read the back of the book, measuring Nature's results. Peeking at the answer in the back of the book, Planck was finagling for all he was worth to get the answer in the back of the book. That's what theoretical physicists do; it is not called cheating. But for all his belief in the continuity, the completion of the limiting procedure which meant that the "atoms" (rungs on the ladder) and their size (rung spacing dependent on the size of h) should not matter, Planck found that the formula with h in it, a *finite* h which had not been allowed to go to zero and disappear from the formula, gave the cutoff and the right shape of the spectrum. A specific value of h resulted in a match of theory with data. Planck announced this in December, 1900, and a nineteenth century self-satisfied classical physics of mechanics, electro-magnetic/optical theory, and thermodynamics, pelted by the radioactive rays in the background, had a fuse now lit by the new theoretical flame.[7] The attempt to unify thermodynamics with electromagnetic/optical theory (as electricity, magnetism, and optics had come together) had produced a new constant of Nature. Particles had been the discontinuities, the indivisible countable things in a continuous world of space and time. Now the continuous fields

7. Planck, *Scientific Autobiography*. 41.

in continuous space and time had brought the world of continuity to an abrupt end: they acted like an energy ladder with a down-side limit to the rung spacing rather than a handicap-accessible ramp. Planck's search for a *universal* science dealing with a *universal* system had produced a surprise: a new *universal* constant.

Planck could not have come close to understanding the total impact of what he had discovered, even though, or maybe *because*, he was a mature man of forty-two. This is unusually old to make a revolutionary discovery in physics, way too old to jump easily to the barricades of a revolutionary new paradigm and call for the ouster of the old regime. But he had worked hard in a system which held promise of universal truths and the invisible ladder with rung spacings had sprung to life; a universal constant had popped out of the blue. He had been neglectful if not dismissive of atoms of matter which could not be seen, but a new atomicity had been forged. He became an anti-Machian convert. He now knew that you cannot dismiss what you cannot see; the metaphysical has real physical consequences. The God he had been seeking in the big (universal systems) and the perfectly accurate (continuity and going to the limit) had come in the still small voice of the small step of a ladder, invisible to the outer eye but necessary to the inner eye, carrying the universal in a very small constant specifying the rung spacings. Einstein's revolution of relativity theory (coming just a few years later) was based on holding the potentially infinite speeds of continuous space-time down to a very large but finite speed limit; Planck's revolution stopped the potentially unlimited division down to the infinitely small by the constant of the smallest rung size. Einstein's theory of special relativity reconceptualized work that had already been done in equations (by others), but his theory of general relativity was a one-man show. He is the undisputed king, founder, and father of relativity. Planck merely lit the fuse of the quantum world; for that reason, I call him the *grand*father of quantum theory.

Planck was quite aware that he had found a pearl of great price. He told his seven-year-old son that his find was to be compared with the work of Newton.[8] If that's what you believe, you can tell a seven-year-old without being pompous or immodest. But he knew and was correct that he had struck something deep in the heart of the universe. Both k and h had been determined by the comparison of his formula with Nature's results, his theory with the ultimate back-of-the-book answer. The constant k was not fundamental; it was simply a conversion from the arbitrary temperature

8. Kuhn, *Black-Body Theory*, 113. Erwin Planck, seven years old in 1900 recalled that, on a memorable walk at that time, his father had said "he had just made the greatest discovery in physics since Newton."

scale to one using the standard units of mechanics. If temperature is basically a measure of hidden energy, why not use standard units of energy, kT, in Joules instead of T, in degrees or Kelvins? Temperature was, in a world of perfect continuity, a measure of an average energy for each piece of Nature, basically anything that could carry energy. Planck's constant h was a new discovery, with hf also being an energy, not an average but the smallest division of energy for a frequency f. When temperatures were lowered or frequency was increased so that the expected average kT (for continuous cases) and the smallest piece hf were comparable, you ran into problems. When a table is many millimeters long, a meter-stick with millimeter markings seems to be a reasonable tool. If the table is expected to be 2.5 mm long, you need a finer measuring stick. And here is the revolutionary piece: when you expect the thermal energy of a given frequency f to be 2.5hf (assuming that the temperature of the whole system of the electromagnetic gas is such that kT = 2.5hf), the energy of that piece *will not, cannot be* 2.5hf. That's between the rungs. It will be on rung 2 (energy of 2hf) or rung 3 (energy of 3hf) *or* it can be on rung 0, 1, 4, 5 . . . and the best we can do is find probabilities. If the average family has 2.5 children, that tells you somewhere to start your guessing but it doesn't tell you anything about a particular family. But you can be sure they do not have 2.5 children. (I sometimes *half*-jokingly say that I have 1.5 Chinese grandchildren because three grandchildren are *half*-Chinese, but this is clearly a different matter.)

When Planck found h, he tore up the plan of perfect continuity, but he completed a trinity of fundamental measures in the universe. Space, time, and causality had been the three basics undergirding Newtonian mechanics, the three legs of the stool of modern science. Space and time were the absolute framework of the stage and performance time for the mechanics drama. Causality had been expressed as forces causing changes in the motion of the actors (the acceleration of masses). We people of faith see skeptical philosophy as destructive of religion. Hume, the great Scottish skeptic, ushered in that kind of attack. But we often forget that he was equally destructive of science. When he convinced the world (as it still seems) that you cannot rationally believe in miracles, he gave equally convincing arguments that you cannot rationally believe in causality. You may see what you call effects regularly occurring after what you call causes, but you cannot make that leap of faith, called causality, with any certainty. Kant, stirred from his slumber by this attack, tried dutifully to save religion and science. Powerful as his influence has been, his saving action is to say that we, as acting subjects (the audience), bring these categories of space, time, and causality to the table and make science possible, just as bringing God to the table makes moral science possible. This he saw as a Copernican revolution in knowledge: the world of knowledge does not revolve around an objective world out there with all the givens built in; the world of knowledge revolves around the subject, the knower. Newtonian mechanics, rightly or wrongly, brought the concepts of length [L], time [T], and mass [M] to use in the basic framework of space, time, and causality. The basic law of mechanics gave force, the universal and *only* cause, a definition: mass x acceleration (ma) or mass x velocity change per unit time ($m\Delta v/\Delta t$). We express this symbolically with brackets [F] and write its result in the fundamental trinity of [M], [L], and [T] as $[ML/T^2]$. To this point in history (1900), there had been two *fundamental* force laws: the universal law of gravitational attraction and the law of electric/magnetic/optical interactions. The law of gravitational attraction was a force law:

$$F = \frac{G\, m_1 m_2}{r^2},$$

F = gravitational force between two objects,
m_1 and m_2 = mass of first and second object,
r = distance between objects,
G = gravitational constant.

In our fundamental terms, the unknown G measured by Cavendish in his famous experiments may be solved for by $[F] = [ML/T^2] = [G][M^2/L^2]$, yielding $[G] = [M][L^3]/[T^2]$. Cavendish found the number; these are the units or what we call dimensions.

The electric/magnetic/optical unification resulted in the speed of light c as a fundamental constant. Its units or dimensions are obviously that of any velocity: $[c] = [L]/[T]$.

Since we have three fundamental units of measure, we need three fundamental constants of Nature to tie them together and to untie them. Planck's constant did this. It has units or dimensions of what we have named *action*. It is easy to decipher from *hf = energy* since frequency is the number of cycles per second and thus has dimensions of [1/T]. Thus, $[h][1/T] = [ML^2/T^2]$. This gives the units or dimensions of h: $[h] = [M][L^2]/[T]$.

Planck saw that these three fundamental constants, all expressed in arbitrary units which we have standardized as kilograms, meters, and seconds, can be used to define a *fundamental* mass, length, and time. These fundamental units are independent of where or when or any arbitrary starting points. They should be the same on any galaxy at any time. (You may recognize that this is a tenet of faith and a bit hard to prove, but physicists have more faith than you think.) The prospects of this universal set of units affirmed his faith in looking for universals and absolutes. These units are called Planck units and represent a wonderful exercise (requiring nothing more than a little algebra) for a budding theoretical physicist like you. Just solve for [T], [L], and [M] in the equations above in terms of [G], [c], and [h]. I don't mind you "cheating" (remember that's what good theoretical physicists do!), so I'll give you the answers here:

Planck time: $T_P = \sqrt{\frac{Gh}{c^5}}$,

Planck length: $L_P = \sqrt{\frac{Gh}{c^3}}$,

Planck mass: $M_P = \sqrt{\frac{hc}{G}}$.

Even experimental physicists would plug in the numbers, and any curious reader might look up the final values. You never know when you might need to know them, and, if you have the faith of a physicist, they will be the same in heaven as here on Earth.

7

Who Will Hear the Still Small Voice?

> A tiny universal absolute constant h had appeared; Einstein saw its meaning in atoms of light, and Bohr was able to use it in his model of the traditional material atom.

> And behold, the LORD passed by,
> And a great and strong wind rent the mountains, and brake in pieces the rocks before the LORD,
> But the LORD was not in the wind:
> And after the wind an earthquake;
> But the LORD was not in the earthquake:
> And after the earthquake a fire
> But the LORD was not in the fire;
> And after the fire
> A still small voice.[1]
> . . . Elijah heard it . . .

IN DECEMBER (AS WE celebrate it) over 2,000 years ago, God spoke softly when he came into the world as a baby born in Bethlehem to a young woman. The hope for a Messiah, a King, a strong man to break the yoke of Roman domination would seem to require a louder voice than angels singing to some shepherds and better connections than a carpenter's family. In spite of the Roman King Herod's fear and the clear uneasiness of the Jewish religious leaders, Jesus's teachings never sounded like a take-over attempt of either the religious or state order. Yet his followers, even the closest ones,

1. 1 Kings 19:11a–12 KJV.

seemed to think that this new Kingdom of God which he preached would set them up in positions of power. While the crowds yelled Hosanna to the King, Jesus walked steadfastly toward the unlikely sounding defeat of death by dying and invited his followers to do the same. No wonder the followers struggled to find the way, even after his resurrection. Being chosen by God should rate us privileges somewhere above suffering and death and give us knowledge a lot more certain than a hoped-for repeat of the most improbably repeatable event in history. The unseen God had been seen in a lowly form and an unseen Kingdom of God had been inaugurated. The oft-repeated theme of an immortality of a soul, reincarnated in a different body with some hoped-for escape from the material body to a spiritual realm of protection, a realm of the gods separated from us mortals, had been crashed by a different theme: a God who chose to come and suffer and die in this mortal flesh, inviting his followers to embrace the material world until he redeemed, resurrected, and renewed it. God had spoken softly before, to Abraham. He said, "Go," and Abraham went. He showed his grace to Abraham's descendants by giving them the law and, in spite of their many failings, they gave this gift to Western civilization.[2] Now God had opened his grace to all peoples with an offer to write the law, fulfill the law, in their hearts if they accepted the gift to follow Jesus *to death*. Baptism had been a sign of repentance and cleansing; now it was a symbol of death to the inevitable failings of mankind in their attempts at the grace of a law covenant; it was a promise of new life. Whether Jew or gentile, coming to this twosome of God/man or this Trinity of Father, Son, and Holy Spirit, to a radical religion of giving your life in promise of a resurrection requires lots of careful consideration, criticism, analysis, and/or brute faith. You are dealing with the invisible, an experiment in the dark with no known criteria. What are the absolutes or are there any absolutes? What can it possibly mean to be God/man and live on both sides of a great divide? It is not surprising that major church councils were trying to sort this out over centuries. A God of the universe who chose to give up the high position, taking the radical choice of letting us choose, and offering us a strange choice is a God who could be almost anything and do almost anything in relation to the universe. You only have one opportunity at this grand experiment, at least in this incarnation, so make the most of it.

In December 1900, Max Planck announced his findings of a law for blackbody radiation distribution, how the energy is distributed over the frequencies in a cavity full of electromagnetic radiation at a given temperature. He had been looking for absoluteness, and certainty, and universality

2. Cahill, *Gifts of the Jews*.

in a system which was independent of the *material* world which we use to contain it. Electromagnetic fields, being everywhere and non-material, had the slight ring of *spirit,* at least a stand-in for the Breath of God, the *wind* and *air* which have served as the root words in many languages for *spirit.* The mechanical and material had been the core of physics and many within and without the world of physics had sought completion and advancement there; many others had rebelled against that mechanism. Darwin had sought mechanism in biology, and Boltzmann had believed in the atomistic mechanics as an alternative explanation or even as a foundation for thermodynamics, becoming a big Darwin fan in the process.[3] Planck considered the world of thermodynamics as a discipline of *principle,* basic laws of systems which were independent of what was *in* the system, whether mechanical or electromagnetic, whether alive or dead, whether discontinuous particles or continuous fields, whether of this world or extra-terrestrial. Just as Jesus said that his words were forever even though heaven and earth pass away,[4] Planck and Einstein thought thermodynamics would not pass away. At least in retrospect, thermodynamics was how the laws of physics *must* be because it was based on the *principle* of the impossibility of the perpetual motion machine. Einstein's great work in relativity was based on how the laws of physics *must* behave if they are to be the same for *all* observers. Thus, both thermodynamics and relativity are the sealing lid on classical physics, the physics of material particles (mechanics) and immaterial fields (electromagnetic theory). This absolute unifying *principle* of how the world *must* be was at the base of Einstein's work and of Planck's attempt at dealing with the thermodynamics of the "gas" of immaterial fields, what modern physics unapologetically refers to as a photon gas.

The laws of physics are for some people just descriptions of what we can experience in Nature. For others, they are the *rulers* of Nature, the rules which Nature *must* observe. Just as the Judeo-Christian tradition posits that there is a higher power who spoke or wrote the laws, scientists in the Western tradition posit that there are rulers of the laws, principles which hold a higher place. This may come from the Judeo-Christian tradition (as Einstein often spoke of God, the Old One, *der Herrgott*, and *der Alte*) or it may come from the Greek Platonic Forms and concepts of an absolute mathematical/logical/philosophical Truth; the two became intertwined during the development of both. Just as God is the Law of Last Resort for

3. Lindley, *Boltzmann's Atom*, 225.

4. Matthew 5:18.

some of us, the laws of thermodynamics are Laws of Last Resort for some (e.g., Peter Atkins, a noted scientist/atheist who wrote a book to that effect).[5]

Planck had the conviction of the absoluteness and authority of the laws of thermodynamics. To say it in theological terms, God provided all energy that Nature needs, energy which can be exchanged but not created or destroyed, and God provided all the order needed, an order which will only be dissipated with time. In secular terms, initial conditions and mechanical laws specify the universe, laws which can be cast in the form of laws of motion or laws of conservation of various quantities. Planck and his friend Ostwald told Boltzmann that his atomic hypothesis which made thermodynamics a branch of mechanics, statistical mechanics, was not necessary. Basically, he believed thermodynamics to rank above mechanics, to be independent of mechanics. And, like many Christians who cannot tolerate any whiff of chance or uncertainty or inaccuracy in God, Planck found the idea that probability crept into his world of absolute laws of thermodynamics repugnant. In the vast numbers dealt with in the atomic or electromagnetic worlds, the possible deviations from exactness and absoluteness were miniscule or extremely improbable. But when we ask the big questions of *principle*, we often do not tolerate *any deviation*. We think we know, or God knows, or the laws know. Planck fell into that trap. But a man of integrity admits when he finds himself in error. Planck did just that.

Einstein was a great admirer of Maxwell who had unified electricity, magnetism, and optics in one fell swoop. Planck was trying to show a similar unifying theme in thermodynamics. But mechanics, specifically the idea of particles which he had hoped to supersede by a higher more general law, reared its ugly head. Einstein was naturally disappointed in the lack of unification. Planck's law was a hybrid, not a unity. One law (Wien's law) worked for high frequencies, one law (Rayleigh-Jeans) worked for low frequencies. Planck had managed to produce a hybrid to which he could ascribe *some* meaning and which he could make plausible, at least to himself. In retrospect, some have considered it a lucky guess and some brought legitimate criticism. He was convinced that the work was highly significant, but his prior convictions probably caused him to miss the greater significance of the still small voice of h, the tiny determinant of the ladder step sizes in what he considered a mathematical trick, and place the emphasis on the Boltzmann constant k. With his emphasis on entropy, he wrote (what Boltzmann had not written) that S (entropy) = k log W, where W is the number of microstates possible for the system to "choose" from in a given macrostate. (A microstate includes the detailed possibilities of all the inner workings, while

5. Atkins, *Four Laws*.

a macrostate only concerns itself with the overall system; macrostates are what classical thermodynamics, Planck's world, is about, while microstates make claims or at least try to see the details.) This definitional equation was considered a fitting triumphant (for Boltzmann) epitaph and appears on Boltzmann's tombstone. The value of k achieved by comparing Planck's formula with the data immediately gave Avogadro's number, the number of atoms in a mole, because it effectively gives the heat capacity of an atom while the heat capacity of a mole had been well-measured. It also gave therefore the charge of an electron because the charge of a mole of electrons was well-known (called Faraday's number). The charge on a single electron was not measured directly for a decade or so. Thus, tremendous progress in physics came from a "lucky" derivation of an obscure system. But the still small voice of h would ring louder and louder as the best minds in the physics world gradually took on the totally unsolved problem of how the electromagnetic world and the world of atoms interacted.

Who were the brilliant physicists available to hear the small voice of h, this new universal constant which seemed to pop out of a "formal" mathematical trick in the attempt to "see" inside the electromagnetic gas? Planck himself said that the way to get new physics accepted, to get revolutionary change, is to have the old generation die off. The old generation can sometimes be convinced that change is necessary, but to develop or accept new ideas requires the youth. Looking at the list of physicists of that era, you can see that we need to look for the teenager or early-twenty-somethings in 1900. There were plenty of mature or grand old physicists in 1900 and their voices counted, but the voice of revolution came from three young Jewish men, Paul Ehrenfest (born 1880), Albert Einstein (born 1878), and Niels Bohr (born 1885). They were very different people, but they were natural matches for each other, becoming close friends, and natural matches for the task at hand. The elder statesman of physics at the time was the Dutch physicist Hendrik Lorentz (born 1853), who shared the Nobel Prize (with Zeeman) in 1902 and did much of the background work for Einstein's relativity theory. Einstein said that Lorentz "meant more than all the others";[6] the common question about any physics topic was: What would Lorentz think? His counterpart in England would be J. J. Thomson (born 1856) who won the Nobel Prize in 1906 for his studies of conduction of electricity in gases. Brilliant and path-finding in their time, both of these men set the stage but were not ready to lead the way into any new promised lands. Henri Poincaré (born 1854), the famous French mathematician/theoretical physicist/philosopher, held similar influence and could add his voice of

6. Heilbron, *Dilemmas*, 25.

encouragement. Ernest Rutherford (born 1871), the transplanted New Zealander, was busy (with the Curies) making and riding the wave of radioactivity and the resulting nuclear science. Big names in German science (aside from Planck), Heinrich Hertz (born 1857) and Hermann Helmholtz (born 1821) had already died in 1894. The legendary Scottish physicist, William Thomson (born 1824), first Baron Kelvin, was a revered "old bird" sleeping through Rutherford's rebuttal of his calculations of the age of the Earth, calculations invalidated by the introduction of radioactivity.[7] The German Wilhelm Wien (born 1864, Nobel Prize 1911), the venerable Lord Rayleigh (born 1842), and Sir James Jeans (born 1877) had done important work on the blackbody radiation, clearing the way for Planck's work, but were unable (or too unbelieving) to extend it. In retrospect, we might think that every good mind in physics would be working madly on radioactivity or the quantum problem in the first decade of the twentieth century. The quantum ideas were in fact very slow to develop and required the dual crises of atomic structure and the relationship of atoms to light. Two decades passed between Planck's 1900 announcement and the first Nobel prizes awarded for quantum physics.

Gaining the title of physicist/philosopher can occur in four ways: 1) destructive criticism, 2) constructive criticism, 3) development of physical theory based on principle, or 4) development of physical theory which brings out criticism and principles. Ernst Mach succeeded too well in his destructive criticism. His limited views of science as an economical summary of sense data led him and Einstein to fruitfully reject the mechanical ether which we could not see. His additional rejection of atoms left him with the rather negative designation of philosopher who had lost touch with reality, an ironic twist of his own negative assessment of all metaphysics. Planck managed to turn his theory based on principles which failed him into a theory desperately searching for principles, criticism, and meaning. Einstein's favored mode of theory was principle-based, exemplified by the theory of relativity. His later work in quantum theory was sporadic, but insightful and influential. He could never find a central principle to clarify his work on quantum physics, frustrating himself and those he criticized, especially his dear friend Niels Bohr. Bohr was more pragmatic in his theoretical work and had great tolerance or even attraction to a fuzzy, even contradictory, philosophy. Ehrenfest contributed almost completely by his clarifying criticism and his personal glue in interacting with the physics community, literally bringing Einstein and Bohr together in his home. The lack of resolution of the difficult concepts, our continuing inability to agree on what we

7. Burchfield, *Lord Kelvin*, 164.

cannot see, almost forces us to consider these men as philosophers, those who ask and hopefully lead us to answers of big questions.

Einstein's greatest work was firmly based in principles such as the constancy of the speed of light and the insistence on laws which were the same for all observers. But Einstein was also an opportunist in his relations with women and with physics. Planck's derivation used quantization of energy of an oscillator, its step-wise nature in only values of 0, hf, 2hf . . . as a mathematical trick. Einstein dared to consider quantization a real limitation placed in Nature, even if we had no understanding of why and how. Planck's *virtual* ladder became Einstein's *real* ladder. That's a real step of faith. Even if that produced results consistent with Nature, we did not know if other paths might equally lead to the same result. The idea that the steps were too big for thermal energy to have much chance of exciting oscillators up the ladder had explained the limitation of the high frequency end of the blackbody radiation distribution; it equally well explained the decrease in the heat capacity of solids when the temperature was too low to excite the vibrational frequencies of atoms in a solid. Atoms in a solid are roughly fixed in a crystalline lattice; their only motion is to vibrate around that fixed position. If each piece of energy has *average* energy of (1/2)kT, the internal energy, that hidden energy which we associate with heat, is easy to calculate. It is just (1/2)NkT where N is the number of pieces of energy. The amount of energy required to raise T by one degree is a constant, (1/2)Nk. But Nature doesn't *always* agree; the erstwhile constant value heads toward zero as the temperature drops below its ability to take the steps up the 0, hf, 2hf . . . ladder. Although Einstein's simple use of this was technically wrong, the basic idea removed a major cloud hanging over thermodynamics.[8] The "old bird," Lord Kelvin, had seen this problem a few years earlier, and this discontinuity of energy, steps instead of handicap-accessible ramps (which classical thermodynamics had assumed), explained a number of such missing terms, apparent pieces of energy which did not show up because thermal energy could not make it up the steps. Big wheels (high temperatures) can make it over bumps (steps of hf), but little wheels (low temperatures) find such a path inaccessible.

Let me emphasize again the problem of a common fallacy in scientific inference. Almost all efforts in theoretical science look into the void where we know and see nothing and make up some hopeful picture. If it works in filling the void, matching with what we *can* see or know, we are encouraged and keep testing in every way we can. But what is sufficient is not always necessary; heat and kindling are sufficient for a fire but there are other ways

8. McFadden, *Quantum Evolution*, 139.

to produce fire. And implications don't always go both ways. If fire always produces smoke, then seeing fire implies the presence of smoke, but seeing smoke does not imply fire. As John Wheeler says, "Where there's smoke, there's smoke."[9] It is hard to be careful in our thinking and presumably philosophers help us do that.

Einstein made one more leap of faith in his *annus mirabilis*, 1905, when he published the special theory of relativity while working as a patent examiner. Rather than considering light in a dark cavity as an oscillator with energies always given by 0, hf, 2hf . . . (a la Planck), he suggested that light came in individual packets, each of energy hf, indivisible, atoms of light. This immediately gave an explanation for a puzzling behavior in what we call the photoelectric effect. Light can knock electrons off of materials, but large amounts of energy at low frequencies are ineffective while a small amount of energy at high frequencies may produce results immediately. If low frequency light has its energy in individual particles of light which are each below a threshold energy required, they cannot combine to produce a single photoelectron emission. Above this threshold, at higher frequencies, each *particle* of light is sufficient to knock one electron off, and emission occurs even if there are few light particles. This provided a convincing natural explanation, and it won Einstein the Nobel Prize (sixteen years later). I do not wish to be a spoilsport, but a critic is right to say there may be other explanations. We do not necessarily believe there are any better explanations, or he would not have won the prize for his work. We may eventually see a different picture.

You would not normally see the work of Paul Ehrenfest discussed at this point. Like Einstein, Ehrenfest was a Jew who bounced around Europe. Ehrenfest was deeply revered and loved by his colleagues, but his work did not gain the recognition it deserved because it was primarily critical. He acknowledged his lack of creativity, his limited mathematical skills, and his lack of clear focus; let's just say it bluntly, he was severely *self*-critical. He loved people, had no pretense, drew no line between his professional and private life (or those of his students), asked all the dumb questions, and had an unerring sense for the fundamental questions. Einstein was from Switzerland and married a Serbian classmate, the only woman in his physics class; Ehrenfest was from Austria and married a Russian mathematician who collaborated with him in much of his work. Einstein played violin; Ehrenfest played piano. They were immediate good friends (only two years different in age), playing music and talking to the wee hours when they were together. Both had trouble getting positions, but Einstein had established

9 Quoted in Bernstein, *Quantum Leaps*, 175.

himself by the time he and Ehrenfest met. Ehrenfest studied in Vienna under Boltzmann and then at Göttingen. Going to Russia, he had good contacts, but he was a Jew, an Austrian Jew. He even had to declare himself of *no* religion in order to marry and live with Tatiana, an Orthodox Christian, but that was not enough to get him an academic position. He seriously considered getting a second doctorate under Sommerfeld at Munich to satisfy a German university; Einstein wanted Ehrenfest to succeed him at Prague, but religion now placed the opposite restriction. The Austro-Hungarian emperor required professors to have and declare a religious affiliation, and Ehrenfest refused although everyone told him it would be a meaningless formal gesture.[10] Then, from Ehrenfest's viewpoint, a miracle occurred. Lorentz, at the height of his career, viewed almost as the king of the physics world, chose to give up his post at the University of Leiden and chose Ehrenfest as his successor. He had first chosen Einstein, but Einstein had just agreed to return to his alma mater, ETH of Zürich; he also might have feared living under the shadow of his hero. Lorentz was a mature cautious master of the whole range of physics and of four languages, Dutch, German, English, and French; Ehrenfest was an impulsive young man who at that time did not know and certainly never mastered the Dutch language. The world of physics at Leiden changed dramatically, never losing the presence of Lorentz who stayed nearby, but gaining in Ehrenfest a brilliant lecturer, teacher, and magnet for great scientists. Building a house patterned after a Russian villa nearby, Ehrenfest hosted students and all the great physicists, collecting signatures of fifteen Nobelists on the wall of his study, which served as a guest-book. With no apologies for no smoking, no alcohol, and no meat, he provided a lively exchange for students and the greatest of physicists, a center for learning, clarification, and criticism.

Planck had fallen victim to belief in the absoluteness and certainty of thermodynamic laws and disliked the thought of any intrinsic uncertainty in Nature, requiring statistics and probability. He probably did not *disbelieve* in atoms but had expected that the use of the concept was unnecessary for the exact laws which must be obeyed. It is certain that he did not have the atomic hypothesis and the counting games of probability theory in his theoretical toolkit. Ehrenfest, on the other hand, worked with Boltzmann, a founder and champion of what became a takeover of thermodynamics by statistical mechanics. Chemistry, the early prod to modern atomism, has found classical thermodynamics (Planck's variety) to be a useful and major part of their curriculum; physics now gives it a light brush in its

10. Klein, *Ehrenfest,* 178. Einstein had registered as an adherent of the "Mosaic creed."

introductory courses and concentrates its thermodynamics considerations in statistical mechanics courses (a la Boltzmann). Boltzmann was asked to write the encyclopedia article on thermodynamics/statistical mechanics (these were book-length treatises, not a few pages). When he committed suicide in 1906, his young student Ehrenfest, not the master classical thermodynamicist Planck, was asked to write it; he did, with the help of his wife Tatiana.

Before Planck's ground-breaking work, Wilhelm Wien had used clever arguments from classical thermodynamics to find properties of the blackbody radiation, winning the Nobel Prize in 1911 for his work. A slow (maintaining equilibrium) adiabatic (no heat flow) compression or expansion of a box of electromagnetic radiation could be treated classically, changing energy (pushing or releasing against radiation pressure) without changing entropy. As concerns of the quantum steps governed by Planck's constant h continued to be primarily in the realm of statistical mechanics/thermodynamics, Ehrenfest continued this work of Wien.[11] He showed that the ratio of the energy in an electromagnetic vibration to the frequency must remain constant in this adiabatic process. The energy increased by work done in squeezing, but the frequency went up in the same proportion, maintaining Planck's ladder of proportionality of energy and frequency. But his clarifying work on counting statistics was to have larger impact, although he generally did not get credit for it. He conceived very simple models of balls being transferred between urns or fleas jumping from one dog to another; these are clearly *particle* models and helped clarify issues between Planck and Einstein. He saw that there were three models for the unseen energies in the electromagnetic radiation: (1) an oscillator at frequency can simply exhibit only energies of 0, hf, 2hf . . . ; (2) the total energy is made of mutually independent units of hf, quantities which may be commingled like a liquid but must always and only change unit values of hf; (3) these units of hf function like atoms of light in absorption and emission *and* have separate identities in empty space, i.e., *particles*. Model (1) was like extension ladders which have a catch at each rung and therefore can have only integral numbers of rungs. Model (3) is like a Lego-set ladder with each rung being a separate piece. Model (2) is like a ladder mold which is filled with a liquid that hardens, the amount of liquid coming only in full-step sizes. Planck did not even accept model (2), and the match with Nature in the blackbody distribution made him look right. Einstein had gone to model (3) to match with Nature in an explanation of the photoelectric effect; thus he also looked right. Ehrenfest showed that Einstein's assumption

11. Klein, *Ehrenfest*, 261.

of *particles* with any kind of particle that we knew would only give Wien's original approximate answer. If these were particles, they had to be fundamentally different, losing their individuality and showing some sort of "correlation which is utterly incomprehensible in classical physics."[12] This was a hint which Einstein was able to follow more than a decade later (in the 1920s), which we will deal with in the next chapter. At this point (1909), Einstein was able to show that the statistical fluctuations in Planck's distribution had a sum of two terms: one term was like independent particles, the counting games a la Boltzmann's atomism, and one term was like interfering waves a la Maxwell's electromagnetic fields. He recognized that some kind of fusion was necessary, particles with non-classical properties like waves and/or waves with non-classical features like particles. Ehrenfest's critical labors on fundamental questions, unheralded but effective, had begun to produce fruit.[13] 1911 is considered a turning point for physics in general and quantum physics in particular, even though no *specific* progress can be tied to that date. A Belgian industrialist Ernest Solvay proposed and funded a by-invitation-only conference on the quantum problem. Planck did not think it would be worthwhile until the theorists *all* felt that their subject had become "intolerable" by the existence of h and that something must be done. Ehrenfest, still seeking a position, was not invited in spite of his significant but little-known contributions. (There was in fact only one invitee who was younger than Ehrenfest.) Poincaré, an elder statesman considered by some to be the last *universal* genius, was invited although he had no experience with the subject. He repeated Ehrenfest's proof that the quantization was necessary for Planck's result, convincing even long-term holdouts, never giving Ehrenfest credit, perhaps never knowing of Ehrenfest's work before dying less than a year later.[14] Lorentz, the chosen and respected chair of the conference, also lent his weight to the arguments and it was agreed that h was here to stay.

One other young physicist, the Danish Niels Bohr, was not invited to the exclusive congress. His mentor, Ernest Rutherford was invited and the Rutherford nuclear atom and its interaction with electromagnetic radiation were to provide the crisis and the opportunity for a major breakthrough. For

12. Klein, *Ehrenfest*, 257.

13. Outside quantum physics, Ehrenfest is sometimes credited with inspiring Einstein's theory of general relativity by pointing out that special relativity makes the rim of a spinning disk shrink while leaving the diameter unchanged. This contradiction with Euclidean geometry, suggested in 1909, gave Einstein the ideas necessary for developing a non-Euclidean geometry of space and time to accommodate gravity in his expanded (general) theory of relativity.

14. Klein, *Ehrenfest*, 251.

about fifty years, spectroscopic analysis had collected enormous amounts of data, literally thousands of measurements of spectral lines, specific wavelengths (and frequencies) associated with the various elements or compounds.[15] One could identify elements by the spectra, literally seeing gold (and many other elements) in the sun and stars. As the electromagnetic theory had developed, there was only one reasonable conclusion: some charged particles must be oscillating or revolving at specific frequencies to produce and/or absorb these radiation "lines" (a single frequency looks like a single line on a spectrograph). When the very light (in weight) electron was discovered in 1897, it was a great candidate for the very rapid wiggles, but no theory presented itself and almost no patterns were found, just raw data, lots of it. Very forbidding.

A full-blown catastrophe had arisen when Rutherford's work with radioactive scattering indicated a very small nucleus which carried almost all the weight of matter, suggesting a simple solar system model for an atom, by now clearly a quite splittable entity. It appeared that electrons acted as planets orbiting their "sun," the nucleus. An astronomical planet can orbit at essentially any radius and has a correspondingly arbitrary period of revolution, its year; more generally, its orbit is an ellipse with the frequency determined by a radius substitute, called the semi-major axis. Something in its past (God?) chose the initial conditions, leaving room for God to make perfect laws that Nature must obey and still have room for variety by choice "in the beginning." In the classic Greek model of the universe, such circular motion was the *natural* state of affairs, not requiring explanation because both the spherical shape and the rotation were considered natural in the celestial realm. In Newtonian mechanics, it was equally natural if God provided the exact gravitational law and the appropriate positions and pushes at some beginning. Thus planetary orbits may have any and all frequencies, God's choice if you are a theist, or even a deist. But atoms somehow must work *only* at very characteristic frequencies, every gold atom having the same set of frequencies. For such variety and complexity as we experience with atoms, there appears little chance at a meaningful model or explanation. No one in their right mind would think they could solve the classical mechanical problem of a number of these electrons going around a nucleus with the electrons repelling each other with forces comparable to the attractive force of the nucleus. In fact, it is a dirty little secret that we physicists have never solved the *three*-body problem exactly, i.e., a sun and *two* planets. We simply do good approximations in our modern astronomical calculations.

15. A recent *Handbook of Chemistry and Physics* lists over 20,000 spectral lines of the elements taken from over 400 referenced sources. Most of these were known by Bohr's time, albeit without as much accuracy as is available today.

And it only got worse, much much much worse. The revolving electrons are supposed to be the source of electromagnetic waves. Following the classical mechanics of the orbit, we can produce any frequency by the choice of a radius (or semi-major axis). But a quick calculation using Maxwell's theory shows that the atom would lose all of its energy in about 100 nanoseconds. And in the process, the electron would spiral in going faster and faster, thus producing higher and higher frequencies, what we would call a light chirp. This is certainly not the well-defined frequencies corresponding to the "pure tones" of the spectral lines. And the natural, vaunted nuclear atoms were equally naturally *unstable*, dying in much less than the blink of an eye. Niels Bohr had the right set of tools, the right philosophical mindset, and an incredibly right set of circumstantial timings to be the man of the hour.

First, he believed Nature over theory, real data over our mental constructions. Theory said such an atom, a classical electrodynamically confined "solar system" of electrons orbiting a nucleus, was unstable. This was clearly not true in Nature. So, step one in the theory:

Step One. *There exist stable atoms.*

There is some manifestation we call electrons; there is some manifestation we call a nucleus; these manifestations both appear together in something we call atoms; we cannot see details of what exists or what is happening inside the space of an atom; our only language and knowledge is that of classical physics, the picture of motion of particles like our solar system. Maybe our language is all wrong, maybe an electron does not exist, maybe a nucleus does not exist, but what we call an atom *does not collapse* immediately and it *can* exist without giving off energy, because it simply *does.*

This sounds a lot like religion. We do not have language to express the ineffable; we cannot see what *is* or what is *happening*; we have a language of space, time, matter, even life and consciousness; we, at least some of us, know there is an experience we call God. We don't know the mysteries involved but there is a sustained and sustaining experience which we cannot ignore or let the theories of others tell us it does not exist. The theory, the practice, the possibilities will *not* exist unless we start with some picture, use what knowledge we have, take a leap of faith and go with it. Electrons do not exist for those who deny their existence. But for those who do believe enough to act, both electrons and God can turn on the light. No man has seen an electron, but we see manifestations which we attribute to an electron.

Now comes step two, a strange mix of our classical world of mechanics and electromagnetism and Planck's strange universal constant h. Planck had started with an electromagnetic oscillator and found that it had to act as though it had energies of integral numbers of hf, f being the frequency. That is, *frequencies* specified the energies, rather than the electromagnetic fields (as in Maxwell's theory). Now Bohr assumed that classical theory of two particles moving in their mutual electromagnetic field gave us the traditional mechanical energy which specified a frequency (by the Planck relationship), a frequency which unfortunately we did not see or know about, a phantom frequency if you like. Most notably, it was *absolutely not* the only frequency we knew about in a classical orbiting system, how many times per second the electron orbited the nucleus. Bohr's rule is simply: find the classical atomic energy, call it for example E_1, and *define* a frequency f_1 by $E_1 = hf_1$. Now suppose there is a series of such "permanent" or "stationary" states, at least not immediately collapsing states. Now we can state step two more completely:

Step Two. *Radiation, the emission or absorption of energy in electromagnetic waves, occurs only in transitions or jumps of the atom from one stationary state to another, and has a frequency given by the difference in the frequencies defined for each state*: $hf = E_2—E_1 = hf_2—hf_1$.

Let's get to the scandal again: neither f_2, f_1, nor f is any frequency we know *classically* from our model of the atom. We do not have any idea that something in the atom is actually moving at any of these frequencies, but f *is* what we have to get, the frequency of light we see coming from the atom, *one* of the thousands of spectral lines measured for decades.

So h has again sneaked into a theory, a theory of a supposed ladder of energies *inside* an atom being converted into frequencies, which is now connected in a strange but simple way to Planck's ladder of energies in electromagnetic oscillators or standing waves. Now all we have to do is have a theory that tells us how one calculates and how one explains the stationary states. How does God keep it from collapsing by giving it a lowest state and how does he choose at what energies the states are? Bohr had a wand that he developed to a magical level: atoms may be weird, following some different rules, but if and when they are *big*, they must give up their weirdness and correspond to the classical world, which in fact had worked quite well, thank you very much, in the *big* world. This he called the *correspondence principle.* Any quantum theory, dominated as it may be by h and its strange ladders with no place to step in between, must have the h disappear and behave classically in the realm of the *big*. Here Bohr's task was to take the simplest system, a hydrogen atom with a single electron presumably orbiting about a presumed nucleus, and find a system of energy levels, various stable orbits, if you must use a classical model, which gets the known radiation frequencies of the hydrogen atom and looks like the classical oscillator for large orbits. That is, the seemingly arbitrary and meaningless difference frequency *must become* the classically meaningful orbital frequency, the frequency of radiation given off *when* the orbit gets big.

Faith, belief in what you are doing, in what is there, is necessary even if the variations from person to person mean we cannot *all* be right. Planck did not believe in Einstein's particles of light. He quite openly criticized Einstein on that point while recommending him for membership in the Academy of Science and for a position at the University of Berlin.[16] Einstein had already earned the right to be wrong. With our hindsight, we see, believe, and teach that Bohr's second step represents a meaningful exchange of energy, the drop in energy in the atom produces a *particle of light* with energy hf. The frequency f_1 and f_2 may be bogus and certainly unrelated to classical frequencies, but the frequency of light coming out, which we can measure, matches the energy given by Planck's formula. Data we can agree on; formulas which relate them reliably we can agree on; what they mean and what exists in the abyss of our unknowing and unseeing give plenty of room for disagreement. Bohr most adamantly did not believe in Einstein's particle of electromagnetic radiation; if you were to prove it today, he would say, you will send the news by telegraph or radio, *electromagnetic waves*, not by particles.[17] But Bohr believed in Nature and in the lawful operation of Nature, even if the abyss of our ignorance lies there all too obviously. And he bridged it with his magic correspondence principle.

Bohr built the bridge from considerations of energy and it seemed magical, contrived, or even ugly to people of different persuasions, which most of the great minds were. This was the opposite of unifying; this was a patchwork Rube Goldberg theory lacking in principle, lacking in clarity; there was not much to love about it. But h had another appearance to make. In retrospect, it seems obvious, but Bohr apparently needed the hint from an otherwise forgotten English astrophysicist named J. W. Nicholson,[18] introducing a new use of h on page 15 of his (Bohr's) lengthy publication. Although it grew in Bohr's work and has become *the* key to presenting Bohr's stationary orbits, he did not even mention it in his Nobel address of

16. Torretti, *Philosophy of Physics*, 315

17. Ibid., 316.

18. Klein, *Ehrenfest*, 275.

1922.[19] I have seen a suggestion that Sommerfeld knew this, but I have not found any reference that predates Bohr's work. We have sometimes called h a *universal* constant because it completed the set of constants to make natural universal units of mass, length, time, (and charge if you include electromagnetism), but it does not readily supply directly the allowed energies because its units are not energy units. To get energy, you need to multiply by frequency f, dimensions of 1/[T], a number which can take on many different values, dependent on the system, certainly not universal. The units of h are the units of *action*, (mass) x (velocity) x (distance), and do suggest a *universal* value of these units, which is the same as the units for what we call *angular momentum*. The mechanics of Kepler's laws can be cast in the form of conservation of energy and conservation of angular momentum. The law stating that equal areas are swept out in equal times is equivalent to conservation of angular momentum: when the distance from the sun to the planet changes, the velocity changes with the reciprocal distance, keeping (mass) x (velocity) x (distance) constant. Closer to the sun, it goes faster; farther from the sun, it goes slower. h is thus a natural step size for angular momentum and a natural size for the lowest step. It would seem at first that this constant angular momentum would not stop the natural expected collapse of the orbit; the electron goes faster as it gets closer, simply spiraling in as *classically* expected. But a new and unexpected magic appears.

Both energy and angular momentum depend on the speed and the radial distance of the electron. But they of course obey *different* formulas. It turns out that if the force of attraction were proportional to $1/r^3$ classical mechanics tells us that circular orbits are mechanically unstable. That is, any disturbance will cause a growing response, the electron flying off or spiraling in.[20] This has nothing to do with the radiation losses; it is purely mechanical. But the force law of attraction is proportional to $1/r^2$ producing stable classical orbits (if we ignore radiation, as we can in astronomical planetary orbits). Specified, quantized, angular momenta provide a condition for specified radii and specified energies. This provided a natural way for God to specify certain states which would be stable. It is as though angular momentum had priority and constrained the atom to stay at certain radii, against the pressures of radiative collapse. (See the Appendix 1 for a little more detail.)

Bohr did not have the advantage of that insight and saw no chance of battling with the massive, seemingly patternless, spectral data. Having returned to Denmark from studying with Rutherford at England's University

19. *Nobelstiftelsen* (Nobel Foundation), *1922–41*.

20. See, for example, Winkel et al., *Instructor's Solutions Manual*, 227.

of Manchester to marry Margrethe and work at the university, a colleague informed him of a simple mathematical relationship (OK, not so *simple*, because it involved differences in the reciprocals of the squares of integers) which a Swiss teacher named Balmer had discovered (in 1885, nearly thirty years before Bohr's work) for a few spectral lines in the hydrogen spectrum. (See Appendix 1.) Bohr says that all became clear when this key, a little order among a mass of data, went into the lock. All that was needed was the quantizing condition on the atom's orbits:

Step Three. *The stationary orbits are fixed by the condition: Angular momentum* = $mvr = nh/2\pi$ *or* $mv2\pi r = nh$, $n =$ *integer.*

With all the appearance of magic, this gave the formula for the Balmer series and also satisfied the correspondence principle, giving the orbital frequency for transitions from $n + 1$ to n at large n. The two uses of h cancelled and the frequency emitted and the orbital frequency, an oscillating charge, coincided, the obvious result expected and found in the classical world. It is still a joy, one hundred years later, to do this algebra and see the puzzle pieces pop into place. All was right with the world. Bohr could sit and wait for his Nobel Prize. (Please know and understand that this is not the way it works. He of course continued the hard work.)

Bohr published this in 1913 in English in very long papers, illustrating the prolixity of Continental writers over the remonstrations of Rutherford, who urged a terse British style. Rutherford did not like the style of writing nor the fact that the electron had no way to know which level it would jump to nor when to jump,[21] but he had to be impressed that the atom suggested by his work had been partially vindicated. At least it might not be all hogwash. Einstein reportedly considered Bohr's work to be "an *enormous achievement*" even though, or maybe *because* such a cobbled mess had produced such remarkable results.[22] Often expressing his views by referring to the deity, he must have felt sorry for a God who made such a mess of things, even if Bohr seemed to read him rightly.[23] At Göttingen, arguably *the* center of the world of mathematics at the time and home to Niels Bohr's brother, noted mathematician and more noted in Denmark as a soccer player in the 1908 Olympics, the famous mathematician Courant was "laughed at for taking [Bohr] seriously."[24] Schrödinger had found Bohr's theory "to be something so *monstrous* . . . really almost *inconceivable*" [Schrödinger's em-

21. Lindley, *Uncertainty*, 53.

22. Klein, *Ehrenfest*, 278.

23. Einstein famously asserted that he had no worries about the correctness of his theory of general relativity. The theory was right; otherwise, he felt "sorry for dear God." Pagels, *Cosmic Code*, 33.

24. Klein, *Ehrenfest*, 275.

phases] because the frequencies did not make any sense.[25] Bohr had perhaps gained some propensity for the understatement of British humor (sorry, I guess that's *humour*) when he admitted it was "in obvious contrast to the ordinary ideas of electrodynamics."[26] It was, to use religious terminology, *absolutely heretical*. Ehrenfest said that Bohr's work "has driven me to despair. If this is the way to reach the goal, I must give up doing physics."[27] Unfortunately, the way to progress in both science and religion often must go down a road of heresy, a tough road for anyone.

This heretical explanation of a few data points in the simplest case among a complex of atomic varieties and thousands of data points could have died a natural death very easily. Could it work for anything else? Well, helium, if it lost one electron in a discharge tube would become a two-particle system just like hydrogen except for the doubled charge on the nucleus. Thus the frequencies should be just double, and we should see a copy of the Balmer series. Aha, it was there but *not exactly* doubled. Is this "strike three and you're out"? No! The extra *mass* of the helium nucleus produced a small correction and Bohr's prediction was right on! From a swinging miss, to a foul-tip, to a solid hit requires only a little correction. A failure becomes a success.

Einstein, as much as he disagreed with style and meaning, probably hit the home run after Bohr got on base. Well before any complete quantum theory came into being, he assumed in 1917 that absorption and emission (Bohr's transitions to higher levels and to lower levels respectively) had equal contributions proportional to the electromagnetic energy at the appropriate frequency in the field (in his mind, the number of light particles).[28] He called these effects *stimulated absorption* and *stimulated emission*. But emission (transitions from higher to lower levels) had an additional contribution due to *spontaneous emission*. No light of that frequency is necessary for an atom to emit, while the presence of light at the transition frequency equally stimulates up or down transitions. No spontaneous emission is possible from the lowest level. (Basically you can have intentional forced steps up or down a ladder (stimulated absorption or stimulated emission) or you can have unintentional (spontaneous) falls *down* the ladder but not up.) Using this simple model based on the Bohr atom, Einstein reproduced Planck's radiation distribution formula, with a relationship between the rates or probabilities of stimulated and spontaneous transitions coming as a natural

25. Quoted in Kragh, *Quantum Generations*, 165.
26. Klein, *Ehrenfest*, 278.
27. Quoted in Segrè, *Faust in Copenhagen*, 110.
28. Einstein et al., *Collected Papers Vol. 6*, 363–70, 381–98.

result.[29] This completed circle of consistency cemented the acceptance of both Planck's and Bohr's works, provided the inspiration for the probability interpretation of QM developed nearly a decade later, and was the basis for the MASER and LASER, which did not appear for about forty years. (The SE in the acronyms stands for *S*timulated *E*mission.) Moreover, the assumption of spontaneity ties the atomic quantum theories to the spontaneity of radioactivity and the ideas of *random chance.* That was a real home run, and by a pinch hitter coming in from another sport. While Bohr had continued his work in the world of atomic theory, Einstein had spent years on a concentrated effort on the general theory of relativity. This is the equivalent of Michael Jordan forsaking his basketball career for a brief shot at baseball and winning the World Series with a walk-off grand slammer. It didn't happen, as you know, for Michael Jordan, but it did for Einstein.

29. Ibid.

8

Improbable Person and Improbable Methods Lead to Crazy New Worlds of Quantum Statistics

> Indistinguishable particles result in statistical phenomena unknown in classical physics, making sense of atom-building and much more.

HE DID NOT GET a Nobel Prize. In fact, he did not get a doctoral degree. He was in the extreme boonies, nowhere near the European bastions of learning. But his timing was great and he parlayed a mistake, some remarkable guesses, and the chutzpah to write right to the top, to a sympathetic Einstein, into one of the most remarkable discoveries of the break that QM had to make with the classical world we all live in.

Satyendra Nath Bose was born on New Year's day of 1894, a full six years before Wolfgang Pauli was born.[1] Pauli had written an extensive, highly praised review of general relativity, a work of hundreds of pages and hundreds of references, by the time he was twenty-one; he had the good fortune to be a student of Sommerfeld, who had been asked to write the encyclopedia article. Bose, writing to Einstein for permission, had *translated* Einstein's general relativity paper into English, a far cry from Pauli's acclaimed work. Pauli, in the thick of the best European centers of science, and Bose, a continent away, were destined to be the respective prods for two

1 And about six years after Ramanujan was born. Ramanujan and Bose represent an interesting study in similarities and contrasts. For Ramanujan's story, see, for example: Kanigel, *Man who Knew Infinity.*

new statistical worlds for mechanics before the full development of QM and its radical descent into an essential statistics for a single individual particle.

Bose was born in Calcutta and had an exceptional record at the University of Calcutta. The oldest child and only son (with six sisters), he received special attention and produced special results, setting a record of 110 percent on the mathematics examination because of multiple solutions. After his master's program, he settled into a teaching career, appointed as Reader at the University of Dacca in 1921. In his early career, he had a few undistinguished publications.

Bose was unhappy with the derivation of Planck's equation for the distribution of radiation. Let's be fair; Planck was also unhappy with the derivation. He had one equation which worked for high frequencies and one which worked for low frequencies; he had cobbled them together in "an act of desperation" to make it work.[2] The fact that it worked and it gave some insight was promising. More than promising, it was revolutionary. Einstein, in 1916 and 1917, had come up with a different derivation, based on the absorption/emission equilibrium with atomic systems.[3] His steadfast insistence on a particle nature of light combined with the cobbled Bohr atom to give new insight and encouragement for both Bohr's "old quantum theory" and the seminal Planck equation. Einstein was sure his ideas were only transitions to a fuller theory, even though we can see in retrospect that the stimulated emission which he posited flowered eventually into the highly successful MASERs and LASERs (about forty years later). As it so often happened, the lifting of the curtain was painfully gradual to the participants, and great insight was required to see where there was progress and where progress might occur. Einstein had masterfully used statistics in 1905 to verify the existence and the size of atoms. By 1917, his use of statistics had extended to his particles of light (in which he seemed to be the only believer), to the process of emission and absorption from Bohr atoms, and to an electromagnetic wave, which would not go away. He thoroughly mastered and fully believed in statistics as an approach to problems of many particles and for complex systems. He did *not* believe it was a fundamental, ineradicable part of Nature. His masterful use of statistics aided and abetted the cause of statistics as an *essential* limitation and core of Nature while he believed it to be a *nonessential* and *transitory* manifestation of our ignorance and the limitations of our calculations. He also knew there was something missing in his derivation of Planck's distribution.

2. Pais, *Subtle is the Lord*, chapter 19, following equation 19.15.
3. Einstein, *Collected Papers Vol. 6*, 363–70, 381–98.

Maxwell and Boltzmann had not waited for the world to know and believe in atoms. Maxwell, already in 1859, had developed a kinetic theory of gases, the forerunner of statistical mechanics. Assuming what is obvious to us now, our almost total ignorance of a seemingly infinite number of particles, he was able to make use of some probabilistic assumptions to find the distribution of molecular speeds. Because the lack of any favored position or direction of velocity yields to an assumption of randomness, and the large numbers allow small enough divisions to approximate calculus, a simple solution can be found by differential equations, the standard technique of mechanics in a continuous space-time. Because the concern is only with relative numbers or ratios, the distribution takes a form independent of an exact size or number of molecules (as long as there are large numbers of very small molecules). The distribution depends on temperature and ties in with well-known macroscopic behaviors of ideal gases. This was thus a window to the connections between the large-scale behavior studied by classical thermodynamics and a statistical treatment covering our necessary ignorance of the detailed structure.

Boltzmann got to the same answer by another means. Maxwell had stuck with the mathematics of Newton, differential calculus, possible because the numbers (of molecules) were so large that small ranges of speed Δv could approximate the infinitesimals dv needed in calculus language and still contain sufficiently large numbers of particles to give good statistical results. Boltzmann took the atoms very seriously, probably feeling a duty to defend his belief against a fairly broad lack of acceptance. He literally thought of the various countable ways that atoms (or molecules or any particles) could be placed (by Nature's thermal shuffling) in various boxes or cells (representing various real possibilities). This reduced the general physical problems to the "simple" (for some people) combinatorics of traditional game and fortune-telling probabilities used for centuries for real macroscopic things such as cards, coin tosses, dice, etc. Our games have the advantages of visualizability, experience, and sufficiently small numbers to develop expertise. For the kinetic theory of a gas which Maxwell had treated, Boltzmann imagined a six-dimensional space, three dimensions of ordinary space (x, y, z) and three dimensions of momentum space (p_x, p_y, p_z) (p is a traditional symbol in physics for momentum). Each particle (in the complete and exact classical view) was designated by its position and its momentum and thus by a point (x, y, z; p_x, p_y, p_z) in this 6-D space. Dividing this "phase space" into cells of equal volumes, he assumed, as a starting point, equal probabilities for a particle to be in any one of these cells. Just as we might cast a die and give it equal probability to have each of the six faces show up, we cast a particle and expect it to "fall" in any of the cells with equal probability. The combinatorial arithmetic, with any constraints such as the numbers of particles and total energy, should allow one to sort out the probabilities. The number of particles in given cells determined the properties of the system; the order and specific identity of individual particles was irrelevant to those properties. Assuming particles are distinguishable, call them particle number 0, 1, 2, 3, . . . , the different configurations would all be possible contributors to the probabilities. Even with small numbers, the number of these various microstates for each macrostate can become large. As examples we will show three microstates (the configurations of the first three columns) all corresponding to one macrostate (the final column with the unmarked particles) of a system with three cells and ten particles. I am calling this macrostate (3, 3, 4), meaning there are three particles in the first cell, three in the second and four in the third. A microstate is one particular way of getting that result; three microstates are shown. For simplicity and ease of viewing, the size of our macrostate is not much bigger that the individual particles, using only ten particles and three cells. In our ordinary world of very small atoms, the numbers of cells and atoms are both huge. If

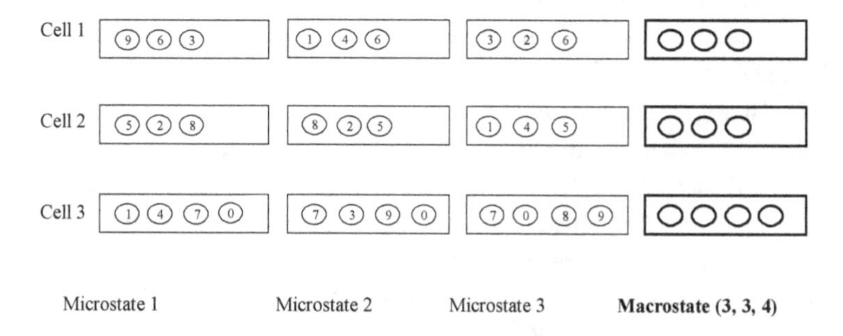

you would like to try your hand at the counting business, you might want to convince yourself that there are 4,200 microstates for this macrostate. It might help you if I write this as 10! /(4! 3! 3!). If it doesn't help you, please ignore it. You can see this gets out of hand quickly. We can barely do a few examples. For example, I found that the macrostate (6, 2, 2) has 1,260 microstates, and with assumptions of equal probabilities for each microstate, (6, 2, 2) has a probability less than a third of that for (4, 3, 3). With mathematical cleverness and approximations for large numbers, Boltzmann was able to reproduce Maxwell's velocity distribution and set a good basis for a statistical mechanics which paralleled the classical thermodynamics.

Before going to a simpler example, let us summarize the rules for what we call Maxwell-Boltzmann statistics. Distinguishable particles are assumed to be randomly distributed into a number of cells or results. Each microstate (distinguishable combination of results) is assumed equally probable. The probability of a macrostate is proportional to the number of microstates corresponding to that macrostate. For example, the toss of a pair of dice gives a total of seven dots by microstates (6, 1), (5, 2), (4, 3), (3, 4), (2, 5), and (1, 6). This gives it a chance of 6 out of 36 or 1/6. Even though this technique can be difficult in our real world applications and was certainly difficult in its application to systems of large numbers of gas molecules, it had been successfully applied in simple cases for several centuries and was a nineteenth-century success that was to usher in the marriage of the classical thermodynamics and statistical mechanics, helping Boltzmann win the battle *for* atomism.

Let us apply this Maxwell-Boltzmann statistics to a simple case well-known long before the lives of Maxwell and Boltzmann. Let us toss two distinguishable coins, make it a dime and a penny. We get four possibilities: (dime heads, penny heads), i.e., (DH, PH), (DH, PT), (DT, PH), and (DT, PT). Therefore, two heads will occur with a probability of 1 in 4. This was probably known as soon as there were coins, maybe as soon as there were cow patties.

But suppose we get two perfectly identical newly minted dimes instead, *indistinguishable*. Now there are only three possibilities: (DH, DH), (DH, DT), and (DT, DT) because (DH, DT) is indistinguishable from (DT, DH). Therefore the probability of two heads has gone to 1/3. If gambling odds can be changed this much this easily, there is a lot of money to be made and many Brooklyn Bridges to be bought and sold. I hope you have objected. This mistake also may have been made as early as there were identical cow patties. I have read (and lost) discussions of this type of mistake in Galileo's time (my memory is that Galileo did not fall for it). This is the type of fortuitous mistake that Bose made in the early 1920s.

You and I know it is a foolish mistake because we have an advantage. We can try the experiment. No matter how hard we try to make two dimes (or cow patties) indistinguishable, we can't pull off the experiment to change the odds. We know that there are two distinct ones because some mark, however hidden, could distinguish them; *besides*, they are *not* in the same space, and God, if not we ourselves, can keep his eye on them. And we are *right*. Don't mess with our common sense.

But God has a better imagination and sense of humor than we do. Because Bose was right, too. He was a bright young man. Let me give you three sufficient excuses for his "mistake": 1) he was dealing with "something" totally unknown, which perhaps only Einstein believed to be particles (it was about this time that Compton's effect (discussed later)was promoting the idea of the particle of light to the larger world); 2) he was dealing with invisible theoretical entities whose "place" could not be verified; and 3) he was dealing even more abstractly with even the concept of "place" or "state," the smallest volume in the abstract phase space to pinpoint a "particle." In other words, he was handicapped by loose concepts which were nevertheless the best understanding of the best minds. His remarkable blundering did much for the development of the concepts needed to make any sense of his (and Planck's and Einstein's) derivation. Wave/particle duality, the uncertainty principle, the complete development of a quantum theory, and elucidation of the symmetry principles involved in quantum statistics were all pieces of the puzzle, and all lay several years in the future.

The story is that Bose made this mistake while lecturing and nevertheless got the right answer. Oh, if it would only happen like that to all of us. That, of course, would only be the beginning of the work. Getting the right answer is nothing. Nature does it every day everywhere, and doesn't seem to care; what matters to us is *why* and *how* it comes to the right answer. Even though the "lecture mistake" may be apocryphal, Bose did not understand the mistake. He understood or sensed enough to know it was important and he put it in the right hands. He tried to publish his derivation in English in the *Philosophical Magazine*. I'm sure they had insufficient understanding to even consider publishing it. (And I'm not knocking the British by saying this.) The young Bengali quietly kicked it up to a higher court, the world-famous Einstein himself, without mentioning the lower court's action, rejection. In a hand-written letter (June, 1924) in English to "Respected sir," he identified himself as a pupil and translator of Einstein's work. He mentioned one specific of his derivation, finding the coefficient $8\pi f^2/c^3$ without using classical electrodynamics, the standard way. (f stands for frequency, c is the speed of light). He wanted Einstein's opinion. If Einstein found it worthy, could he please get it published in German? It set Einstein's wheels churning, distracting him briefly from his search for a unified field theory and his secretarial affairs.[4] Einstein was impressed, returning the translation favors by personally submitting it for publication in German with his personal seal of approval attached to the end. Moreover, he immediately set to work on some extensions, using the new ideas to give a quantum treatment of an atomic gas.

These papers of Einstein written in 1924, well before the formulation of QM, are mind-blowing, suggesting the potential of what he might have been able to do in QM if he had put himself into it. In spite of his statement that he had spent 100 times more time thinking about quantum physics

4. Parker, *Einstein, Passions of a Scientist*, 205.

than about general relativity, his incursions into quantum physics seemed more like interludes. My own opinion is that he insisted on a cleanness and clarity in his work, something he never found in the quantum world, or found only in small pockets. The interludes of this time including his encouragement of Louis de Broglie, as well as Bose, were to produce his last constructive contributions to QM.

Einstein's papers of 1924–25 were masterful and predicted a phase transition in which molecules (ones which behave in this way, i.e., like the photons of Bose's photon gas) would "condense" at low temperatures to the lowest level, accumulating to the single state of zero velocity. This came to be called Bose-Einstein condensation and such particles came to be called bosons. Beautifully crafted theory does not guarantee truth. Einstein, in a letter to Ehrenfest, says that "the theory is pretty, but is there also some truth in it?"[5] It was quickly accepted in the canons of physics, but it required seventy years before its experimental actualization. By then, of course, Einstein was long dead (for forty years), even the younger Bose (for twenty years). These papers would not rank in Einstein's top five, but would clearly assure Einstein of greatness if they were the only thing he had written. This work, clearly initiated by Bose, also immortalized his name, and assured him a secure place in academia in India when accompanied by Einstein's recommendation.

Bose's paper had been a bare-bones derivation of Planck's radiation distribution, despite containing astonishing new ideas. The lack of comment on these features gives credence to the legend of the mistake. He later confessed that he had not seen the significance of his own work. There were three novel features in Bose's work: 1) Even though he treated light as particles, the number of particles was not conserved or fixed. Photons could be freely created and annihilated in exchanges with the wall of the container. In a sweet irony, the word *photon* was coined and first used in a 1926 Gilbert Lewis paper titled "The *Conservation* of Photons" (my emphasis), claiming the photon was the "uncreatable and indestructible" atom of light. A readily accepted name, wrong concept.[6] 2) The *distinguishability* of particles assumed by Boltzmann is no more. 3) Statistical independence of particles is also gone, because of their indistinguishability. Bose got all these right, although his lack of comment makes one believe he might not have noticed. He seems to have felt that he was doing what Boltzmann would have done. Abraham Pais opines: "I believe there has been no such successful shot in the dark since Planck introduced the quantum in 1900."[7] Even with its nega-

5. Pais, *Subtle Is the Lord*, 432.

6. Ibid., 407.

7. Ibid., 428.

tive tone, this is high praise and good company for an otherwise unknown Bengali whose musical talents on the Indian equivalent of the violin would have placed him musically with Einstein and Planck and whose multilinguistic talents set him a notch above.

Ehrenfest and others noted and objected to the lack of statistical independence. Einstein with his prediction of condensation was fully aware even if mystified by it. The indistinguishability of particles and the resultant difference in counting procedures between the Boltzmann and the Bose-Einstein papers "express indirectly . . . a mutual influence of the molecules . . . of a quite mysterious nature."[8] Rather than the ancient action-at-a-distance problem, we have a new problem: these particles, contrary to our commonsense material chunks, can occupy the same space, the same state, but they "see" and "feel" each other; they affect each other's statistics by the radical indistinguishability of particles. The fact that an exchange of two of them produced no change means that there are real statistical effects as if exchanges really occurred. With these particles, the odds of two heads or two tails really would be changed from (1/4) to (1/3). The hypothetical exchange, if that's what it is, really does produce effects as though there were "attractive forces." A head attracts a head making HH more likely than in the Maxwell-Boltzmann statistics.

This also solved an obscure problem of thermodynamics. The third law of thermodynamics can be stated this way: as the temperature goes to absolute zero, a system settles down to a single state (perfectly ordered and therefore with zero entropy). Nernst had dared in 1914 to suggest that this should apply to gases as well as solids and liquids. Einstein, now in 1925, pointed out that a Boltzmann gas would not satisfy this law, but that a Bose-Einstein gas does. One might have the audacity to suggest that there *does not exist* such a thing as a Maxwell-Boltzmann particle; i.e., no basic particles are in fact distinguishable. Of course, I do not mean that an electron is not distinguishable from a proton. I mean that any two protons (or any two electrons) are radically indistinguishable.

Pauli set the ball rolling for the *other* statistics in 1925 in what might seem to be an unrelated way. Electrons seemed to march into atoms a bit like animals into Noah's ark, two by two. Pauli solved this conundrum by a clever assumption which eventually earned him the Nobel Prize twenty years later. Generalizing our commonsense understanding that two particles cannot occupy the same space, Pauli suggested that no two electrons can occupy the same *state*. The counting of states was a matter of some subtlety, one with which Bose had successfully (if accidentally) dealt with in the matter of

8. Ibid., 430.

photons freely roaming in a volume. In those circumstances, a *state* simply had a certain volume in phase space (x, y, z; p_x, p_y, p_z) of h^3. Well, even there we see a funny intrusion of a factor of 2 which comes in because of two different polarization states of light. This factor of 2 was necessary but had no effect on the *relative* distributions. *States* in atoms had been counted by integers from the old Bohr quantum atom. By this time, the periodic table and some basic understanding had given good indications of a shell structure: shells of 2, 8, 18, etc. electrons. This was most dramatically seen by the chemical inertness of element #2, #10, #18, #36 (helium, neon, argon, krypton) as though shells (or subshells) of 2, 8, 8, 18, 18 . . . had been filled; equally clear was the high activity of elements with one missing in a shell (fluorine, chlorine, bromine, iodine, i.e., elements #9, #17, #35, #53) or one extra outside a shell (lithium, sodium, potassium, rubidium, cesium, i.e., elements #3, #11, #19, #37, #55). The numbers (for closed shells) are much more recognizable if we divide by 2 and obtain 1, 4, 9, etc., i.e., 1^2, 2^2, 3^2, etc. Clearly this showed promise to any Pythagorean mystic or numerologist who can rejoice in a return to simple whole number arithmetic. Back in the origins of kinematics, Galileo had noted that distances of 1, 3, 5, etc. in each successive time interval had added up to 1, 4, 9, 16 (a quadratic relation of total distance with time, typical of a falling object or an object going down an inclined plane). If we assign a main quantum number n the possible values of 1, 2, 3 . . . a secondary quantum number l to be 0, 1, . . . n—1, and a tertiary quantum number, m, to go from $-l$ to l, i.e., $-l$, $-l + 1$, . . . , $+l$, we have produced a numerical framework almost matching the periodic table:

$n = 1$	$l = 0, m = 0$	1 state
$n = 2$	$l = 0, m = 0$ $l = 1, m = -1$ $l = 1, m = 0$ $l = 1, m = +1$	4 states
$n = 3$	$l = 0, m = 0$ $l = 1, m = -1$ $l = 1, m = 0$ $l = 1, m = +1$ $l = 2, m = -2$ $l = 2, m = -1$ $l = 2, m = 0$ $l = 2, m = +1$ $l = 2, m = +2$	9 states

Going back to Bohr and Sommerfeld, these numbers were "understood" schematically as *n* producing different size orbits with different energies, *l* labeling different allowed angular momenta, and *m* designating different possible orientations of the orbits.

Pauli's contribution and principle can be stated fairly easily. It involved the factor 2 and the orderly filling of the shells. To get the factor 2, he suggests an additional set of quantum numbers which partially ruins the pure parade of integers. He suggests that the electron itself has an intrinsic angular momentum, as though it were a spinning top, given by a half-integer $s = (1/2)$ (*s* stands for "spin") and the ladder of possible orientations has only two values, $m_s = +(1/2)$ and $m_s = -(1/2)$; no comment here about which is male and which is female. This may seem like a puny contribution to win the Nobel Prize, but Pauli was always in the thick of developments and debates. It was probably unthinkable that someone so involved would not get a Nobel Prize; one can think, however, of others who were highly regarded and deeply involved in the quantum developments who did not get the prize (Sommerfeld, Ehrenfest, and Kramers, to name a few).

Pauli had hit upon a secret which was the complement of Bose's contribution: the world consists of two classes of particles, particles that were indistinguishable from their compatriots and could share states (e.g., Bose's photons) and particles that were indistinguishable from their compatriots and would not share states (e.g., Pauli's electrons). Einstein developed the statistics for the entire first category, which we call bosons. Fermi and Dirac developed the statistics for the second category, which we call fermions. Quantum statistics includes only these two categories; no other type of fundamental particle exists (to our knowledge).

Let us return to the simple statistics of a system small enough to do our counting easily: two "coins," each with three sides, heads, tails, and feet.

Maxwell-Boltzmann considered distinguishable particles, call them A and B. There are nine two-coin states, A with three different states and B with three different states.

State	Heads	Tails	Feet
1	A		B
2	A	B	
3	AB		
4		A	B
5		AB	
6	B	A	

7			AB
8		B	A
9	B		A

State 1 has coin A coming up heads and coin B comes up feet, etc. Three of these nine have the two coins with the same result, state 1 with two heads, state 5 with two tails, and state 7 with two feet. Therefore, there are 1/3 of them in the same single-coin state.

Bose-Einstein coins are indistinguishable, so we will call them both A. Now there are only six distinguishable two-coin states.

State	Heads	Tails	Feet
1	A		A
2	A	A	
3	AA		
4		A	A
5		AA	
6			AA

Half of these states have coins in the same state, two heads, two tails, and two feet. Therefore, there is (1/2) chance of the coins being in the same state, considerably higher than the chance for Maxwell-Boltzmann coins.

Fermi-Dirac coins are indistinguishable, but they cannot go into the same state. It seems superfluous to write down the only three possibilities.

State	Heads	Tails	Feet
1	A		A
2	A	A	
3		A	A

Of course, the chance of two coins being in the same state is zero; they *cannot* both be in the same state.

A comment on each might now be helpful. In Nature, *no fundamental* particles are Maxwell-Boltzmann even though the complex *non*-fundamental particles are what we experience in coin tosses, dice throwing, etc. Moreover, even in microscopic, fundamental Nature, Maxwell and Boltzmann did statistics *almost* correctly on particles which were simple enough to be indistinguishable. When there are *many* particles (so that statistics works well) and *many more* states for the particles to fit in, the statistical distribution is independent of whether only one particle is allowed per state

(Fermi-Dirac) or many are allowed (Bose-Einstein). This occurs when the average occupancy level (average number of particles in a given state) is much less than 1. Multiple occupancy is then so rare that statistics which prohibit it (Fermi-Dirac) or enhance it (Bose-Einstein) have essentially no effect. This then is what we call the *classical* statistical limit: the ratio of particles to available states is much less than 1. Then all statistics are approximately the same and correspond to Boltzmann's analysis. Effectively, the indistinguishable particles act like distinguishable particles because they rarely are in the same state.

One more comment. We probably all know that the number of particles is absolutely huge. A mole of material is of reasonable size, something to be held in the hand if it is solid or liquid, in an ordinary balloon if it is a gas. A mole contains 6×10^{23} or 600,000,000,000,000,000,000,000 molecules, unfathomably many; a million of us could count them in about the age of the universe, 10 billion years. We learned this number only in the early twentieth century. How do we count the available states? Even though we did not understand it immediately, Planck's discovery of the constant h named after him in 1900 was the key; Bose's naïve calculation in 1924 and Einstein's keen perception was a key breakthrough in understanding it. In principle, there is no limit to the number of states available. h is made up of the product of *length* and *momentum*, each of which exists in three dimensions, making a *size of a state* to be h^3 in our 6-D phase space of *real space* plus *momentum space*. For a batch of particles under consideration, real space will be confined, e.g., the size of our container. Momentum space could in principle be as large as you like; in practice it is limited by temperature which gives an average energy. To get the number of available states at a given temperature, we simply take the product of volume of the real space and the volume of the momentum space at that temperature, divide by h^3, the *size* of an elementary state. As you can imagine, doing this yields numbers much greater than the number of particles for ordinary circumstances like a gas at standard temperature and pressure. The condition then for classical statistics, when everything behaves as Boltzmann imagined, can be stated simply: a sufficiently high temperature so that the number of states available greatly exceeds the number of particles. To behave classically, the world has to be thin in particles and/or high in temperature. (We sometimes recognize a budding experimental physicist by their play with physics toys and a budding theoretical physicist by their play with numbers. If you are just dying to try the numbers to find out if the *universe* acts as a classical Maxwell-Boltzmann or a quantum system, you are a prospective theoretical physicist/cosmologist. I won't spoil it for you, but I will give you the estimated number of particles to be something like 10^{80}. Actually,

I probably should not give this number because it only includes electrons, protons, and neutrons; there are estimated to be one billion times as many photons, i.e., 10^{89}. See if that changes your results. This is clearly an extra credit problem.) In effect, particles *appear* to be distinguishable when there are *relatively* few of them, and this clearly happens in much of our everyday world of experience.

When a system cools down, the number of available states goes down and the true nature of the particles, whether they are bosons or fermions, starts to make itself felt. We call such a system "degenerate," which effectively means we cannot continue to use classical statistics but must start to see quantum effects. In spite of the sensitivity to the e-word, the cooling and condensation and fall into degeneracy are a big part of the cosmic *e*volution process. If you dislike the e-word, you may be comforted by its association with degeneracy. It has become fashionable in some biological circles to try to fight evolutionary ideas with the concept of "irreducible complexity." The idea is that *Nature* is not clever enough to produce some level of complexity on her own and that "supernature," an "intelligent designer" is a necessary addition. If God is creator of all, this simply makes him into a two-stage creator and the division between the first stage, Nature, and the second stage, Supernature becomes completely unimportant; I believe it is not an overstatement to suggest this is a meaningless distinction. The evolution of a richly varied physical reality based on such simple principles (the quantum statistics above) makes me want to suggest that we might better find God's amazing work of biological reality in an *utter simplicity* rather than the complex. In any case, the physical world as we experience it can be viewed as very few kinds of particles settling into necessary or highly probable configurations as the universe cools due to its expansion, a well-known phenomenon at the scale we can experience.

"Fermionic condensation" (nobody calls it that) dominates our landscape for two reasons: 1) fermions have the strict prohibition against occupying the same state, giving them an effective strong repulsion, and 2) our two common types of fermions are given also a strong attraction for each other by having opposite electric charges (protons and electrons). It is no accident that the first quantum system to start getting a specific model was the hydrogen atom where an electron gets pulled into a space (near a proton) that is small enough to be the only state available to it. In such small confines and in solids, anything close to room temperature is low temperature and the condensation consists of each electron falling into the lowest state available to it. Basically all of what we may call chemistry and thus the kickable materials within kicking range is concerned with each electron finding a place of its own. All of the common properties of the material world around us result from the compromise in the tug of war in which electrons are pulled toward positive charges but refuse to share the same state. The anti-social nature of the fermions keeps our physical world from collapsing. It gives our world shape. To restate our concern of the previous paragraph, the *simplicity* of the Pauli exclusion principle or Fermi-Dirac statistics resulting in fermionic condensation is the way God produced the amazing complexity of *all* chemistry.

The strict prohibitions of the anti-social fermions are not subtle; they dominate what we think of as the material world. They actually produce *kickability* by not allowing the electrons in my toe to penetrate the space taken by other electrons. The bosons, who are sociable, not only allowing but preferring to share states, are much more subtle. They are more like the bath water, the air we breathe, and the fillers of our empty spaces. The subtlety of bosons makes it remarkable that the breakthrough into the quantum world, just a bare crack, not a model, came in Planck's treatment of what would eventually come to be thought of as a photon gas, a box full of bosons. The reason the subtlety could slip by unnoticed before Planck was that thermodynamics was a powerful tool for dealing with the thermodynamic equilibrium of the photon gas (or electromagnetic radiation in a cavity, as it was thought of then) with its material container, largely independent of the details. The reason it did not fit in and was discovered was that it was a *strange* gas: the number of particles was not fixed and they obeyed a strange statistics, later exposed and labeled as Bose-Einstein statistics. Coming from the thermodynamics point of view, Planck was able to get the right equation and leave enough uncovered for the new features to eventually be seen. He did this in spite of not believing they were particles and therefore not having a clue that the number was not fixed or that they could have any new form of statistics.

When Bose made the mistake of deriving photon statistics with a set of strange assumptions, Einstein immediately applied them to *material* systems which did have (presumably) a fixed number of particles. The photon gas was degenerate, i.e., unavoidably obeying the quantum statistics of what came to be called Bose-Einstein statistics. It does exhibit the social nature, the greater likelihood to share the same state than its classical counterparts, but as lowered temperature lowered the number of available states, it also lowered the number of particles and thus scaled down without doing anything dramatic. Einstein, with the sure-footedness of a mountain goat in difficult terrain and the insight to spot new paths, immediately saw the possible consequences if *material* particles behaved as the photons appeared to. The additional constraint of a fixed number of particles meant that such a gas could not simply scale down at lower temperatures. It meant that such particles, allowed to be in the same state against our common sense, would statistically prefer that and would condense, all huddling together in *one* state when they got cold. Even though it took seventy years to achieve the Bose-Einstein condensation of a gas, the effects of the gregarious nature of bosons produced many features of cold bosonic materials, such as superfluidity and superconductivity. As it turned out, all particles with half-integral spins (intrinsic angular momentum) were fermions and any even number of fermions forced to stay together produced a composite with integral spin, which was the condition to be a boson. Thus ^{4}He, the most common helium atom, has two protons, two neutrons, and two electrons and is thus a boson, and exhibits superfluidity. The simple improvement of odds which we calculated for two heads or two tails in a two-coin toss takes on much stronger forms as the number of particles increases and we go to low temperatures.

In short, the marvelous variety produced at the chemical level, making chemistry a separate discipline which is too difficult for a physicist, is all produced by a few simple particles and these simple rules of relationship. You can call it *natural* or *supernatural*, but it does make it easier for a physicist to believe that amazing complexity can come from very simple rules. If you don't believe or want to bother with a God who might make, sustain, and enforce the simple rules, you think *Nature* is remarkable. If you believe in a creating, sustaining God who did it, you think *he* is amazing. As he often does, he gives you that *choice*.

Let's backtrack a bit to try to clarify the intrusion of statistics. Einstein could be such a bear in his objections to probabilistic thinking, and yet he was a prime mover, with Bose's almost inadvertent prodding, in starting the field of quantum statistics. Statistics and the calculation of probabilities were traditionally very clever shaping of our ignorance. Maxwell had done the first shaping in the spirit of classical physics, with differential equations. Thermodynamics had cleverly taken the *black box* approach and developed laws independently of the details inside. There was no shame in admission of ignorance and a statistical model gave an additional way of looking at it. Boltzmann reproduced Maxwell's work with the conviction of atomism and a conviction that we could never handle the details. With that impetus, statistical mechanics and thermodynamics were fated to become one, giving equivalent results whether you do details by statistics or cleverly avoid details altogether. At that stage, nobody doubted that the details followed the mechanical determinism and distinguishability inherent in a system which *in principle* allowed you to follow every particle exactly. Ignorance came as a result of the large numbers of particles and our real *practical* limitations in measurements. None of us had the super-intelligence of Laplace's demonic master of the universe, but such possibilities were not in *principle* ruled out.

Ignorance took a big turn to the worse with the fundamental indistinguishability of particles. We cannot follow large numbers of particles and, practically speaking, we can *never* follow them exactly; now we add the real possibility that there are particle exchanges which we cannot know, which we can *never* follow in *principle.* This is the problem of dating a twin, but multiplied by 10^{23}. It has been suggested, only half-jokingly by Wheeler, always creative, clever, and controversial, that there may be only one electron in the universe but it just runs around and plays the part in every occupied state. Besides this problem of indistinguishable particles trading places, the ease of creating or annihilating photons meant that the number of particles was by no means constant. Einstein made no objections to statistical usage in this case; in fact he made most masterful use of it. He was not embarrassed by his or our ignorance.

Einstein *was* embarrassed by God's ignorance: "God does not play dice," he famously claimed. Or to leave the God-language aside, Nature *surely* has perfect law and order. When we get down to *one* particle or the tossing of *one* die, we may admit our *practical* limitation. We cannot predict the result of a roll. That's why we use it for our *games* of *chance.* It is *chance* to us because of practical limitations and we *play* with it. But physics is not a game. This is our reality, God. You can't gamble with us. Einstein was the Calvinistic non-believer. His impersonal God is not allowed to play craps with his creation. Einstein and R. C. Sproul might get along. (R. C. Sproul wrote a book called *Not a Chance,* which gives the Christian version of Einstein's aversion to fundamental chance.) It is hard to let a well-ordered God be personal, whether you are Christian or non-Christian.

An essential inextricable probabilism dealing with an individual particle was something that stuck in Einstein's craw. We find assertions by Einstein and friends who declare that Einstein was not insisting on determinism. Maybe Einstein saw other ramifications of the uncertainty which he could not swallow. Certainly uncertainty and a probabilistic interpretation were unpalatable. He kept taking the seemingly inevitable steps right up to the door and then balked, never willing to cross that threshold into the la-la world of QM. Wave/particle duality, the minimum size of a state in phase space, the uncertainty principle and Born's interpretation of ψ as a measure of probability (all treated later) just kept marching on, but Einstein wasn't in the parade.

From 1905 to 1923, Einstein was alone in his belief in the light-quantum. Thus he was doubly alone and far ahead of his times in accepting a duality of something he alone believed in with the only thing the rest of the world believed in. In 1909, he expressed the opinion that "the next phase . . . will bring us a theory of light . . . as a kind of fusion of the wave and the emission theory [particle view]." Planck had mildly and diplomatically renounced Einstein's particle view of light suggesting that risk-takers cannot always be right. Millikan, in doing the Nobel-winning work of verifying Einstein's results of the photo-electric effect, had nevertheless declared "the (Einstein) semi-corpuscular theory" to be "wholly untenable," softened only by the weasel words: "[it] seems at present." Einstein's early (1909) anticipation of wave/particle duality in light and his later encouragement of de Broglie's extension of that duality into the matter world set the stage, raised the curtain, and turned on the spotlight for Schrödinger's full-fledged development of QM.

Bose and then Einstein had made full use of Planck's h as a minimum size in phase space for a single state. Heisenberg's uncertainty principle was just a strong statement of that same principle. Similarly, Einstein's papers of 1916–17 which derived Planck's law inspired Born's interpretation of as a particle probability; Born was mimicking Einstein's use of the square of the electric field as a probability of a photon.

Our use of small numbers in classical devices, coins and dice, to illustrate quantum statistics may obscure the fact that two levels of statistics are at work. The statistics of large numbers of identical particles continued to be supported by Einstein and was immensely fruitful. The unavoidable probabilism of individual events was rejected by Einstein but was no less fruitful.

One might reduce our basic physical mysteries to two principles: 1) the constituents of the universe can act in two distinguishable, even apparently contradictory ways, as waves and particles; 2) when they are acting as particles, or seen as particles, the particles in a class are absolutely indistinguishable. Mysterious but simple; subtle but hopefully not malicious. Certainly a long way from the impersonal machine of deterministic classical physics; hopefully worthy of a personal God.

Let me tell you a cautionary tale. James Clerk Maxwell was a wonderful Church of Scotland elder, and arguably the Einstein of the nineteenth century, i.e., the foremost theoretician in physics. He succeeded in the unified field theory of his time, unifying electricity, magnetism, and optics. I already told you my biggest and most expensive textbook in college was a massive tome which we called the bible of optics. The first page contained Maxwell's crystallization of electromagnetic theory, the four Maxwell equations. The next 1,000 pages work out their results in optics. Einstein considered Maxwell a revolutionary who ushered in "a change in the conception of reality [to a world of continuous fields] . . . the most profound and most fruitful that physics has experienced since the time of Newton." As we have indicated, Maxwell was also an early believer and contributor to atomism in his work on the statistical distribution of speed in the kinetic theory of gases. His belief in atoms clearly took on a metaphysical meaning quite uncalled for by the physics. Despite the course of history which involves catastrophes and evolution, he believed that atoms were forever: "unbroken and unworn. They continue this day as they were created perfect in number and measure and weight . . ." This clearly hearkens to creation and a Creator, but just as clearly does *not* hearken to a biblical account which has *history* and *change*. It is easy for the greatest and best motivated people to get it all wrong sometimes. Faraday, his experimental counterpart among nineteenth century physics greats and a Christian of equally good credentials, took an agnostic course with respect to atoms, declaring the atomic doctrine to be an *assumption* of truth, maybe an article of faith, maybe a working hypothesis, certainly not a clearly predetermined known.

When one begins to glimpse the limited number of fundamental particles and the marvelous array of all the elements and all the chemistry which results in everything we call material reality, it is indeed tempting to suggest both their special creation (even though the Bible would not speak of atoms at all) and the intelligent design indicated by the irreducible complexity and the low probability of such construction. And yet from these few basic principles, a story of *natural creation* of all the elements and their properties has emerged convincingly. And *yet* again, these few basic principles came from somewhere, and this story of natural creation neither eliminated God nor elucidated him in any essential manner. Statements we can make about God from such glimpses sound puny at best. Questions about whether God plays dice may be substantive; to suggest God is a weak left-hander (as Pauli did in the 1950s) is probably not helpful.

9

Judenphysik to Knabenphysik

The very young Heisenberg and younger Dirac finally produce a formal theory of QM

WITHIN THREE YEARS OF Planck's 1900 announcement in the still small voice of the tiny universal constant h, a baby boy was born, a boy destined to bring the volume up to a roar, a trumpet blast, of a new era in which h held a central place, dare I say a central throne. The grandparents and great uncles of quantum physics were not to die off quickly, but the fathers of quantum physics, the young men who could relish and jump on new ideas were just then babes-in-arms. In fact, three baby boys were born in those three years, destined to give the new physics the moniker *Knabenphysik* (boy-physics). Seeing two of the three arrive unmarried, with their mothers in tow (or vice versa), in 1933 in Stockholm to receive Nobel Prizes in physics shows how far and fast the new generation advanced. The three toddlers of the beginning of the twentieth century were: Werner Heisenberg (1901–76), Wolfgang Pauli (1900–1958), and Paul Dirac (1902–84). Heisenberg was the Great White Hope of Germany; Pauli an Austrian Jew from Vienna; Dirac an English engineer born of a Swiss father and English mother. They were a most unlikely trio: an athletic (and musical) German Boy Scout (*Pfadenfinder*, path-finder, he was called), a brilliant Jewish mathematical prodigy who tended toward a teen-ager's stay-up-late-sleep-late lifestyle, and a quiet tall English loner who was most at home on a long lone walk.

Pauli was Wolfgang Jr, living in the shadow of a distinguished biochemist father, professor at the University of Vienna. Until a few years before young Wolfi's birth, the family name was Pascheles; the grandfather had been a stalwart member of the strong Jewish community in Prague. Wolfgang Sr had converted to Catholicism and changed his name, presumably widening his opportunities in the anti-semitic Europe in which Jews appeared to still be wandering in the wilderness, strangers in a strange land. He had been a friend of the son of Ernst Mach, the famous anti-metaphysical philosopher/physicist at the University of Vienna. Wolfgang Jr was christened Wolfgang *Ernst* Pauli when Mach agreed to be his godfather. Mach was in his dotage (dying in 1916) by the time young Pauli was ready for university studies. Boltzmann, also a physicist at the University of Vienna, had committed suicide in 1906, and the home-town university did not seem the great place for a great young physicist to study. Arnold Sommerfeld (1868–1951), the unquestionable super-teacher of twentieth-century physics, thus gained one of his many star pupils at the University of Munich (*München*). His hall-of-fame list of students and his eighty-four nominations for the Nobel Prize (more than any other physicist) speak for themselves. Einstein passed the honor of writing the theory of relativity encyclopedia article (a book-length compendium) to Sommerfeld, who proposed to collaborate with his freshman physics prodigy Pauli but then gave him full responsibility as he realized his capabilities.

Even though this was only a few years after Einstein's general relativity was published, the young man's work earned Einstein's high praise and was published as the classic long-standing exposition by the time Pauli was twenty-one. It is no wonder that Sommerfeld could abide the young man's habit of sleeping past class-time, and everyone (except perhaps Dirac) could abide the seemingly unerring criticism from his sharp-sounding tongue. A simple honest know-it-all who really seems to know it all was a great asset to the physics community. Muhammad Ali could get by with "I am the greatest," at least among his own community. So could Pauli, although he did not say it that bluntly. Ehrenfest and Pauli, both Viennese Jews, could start out as fast friends, trash-talking immediately: "I like you better than your work." "Oh, with me, it's just the opposite." It doesn't matter who insulted who first, or how. Pauli, known as the Scourge of God and the conscience of physics, was clearly loved by the physics community and valued for his judgment, even when it was damning. He may not have always been right, but you should probably not bet against him. (In his later years (his fifties), he did bet in favor of the ambidextrous universe, a universe equivalent to its mirror image, and *lost*, admitting that God was, in his words, a weak left-hander.)

Heisenberg would have been an Eagle Scout who played baseball, eating hot-dogs and Mom's apple pie if he had been American. But he was the German son of a classics professor and more German and more a musician than physicist (according to some), and deeply involved in the youth movement. In the jumbled mess of the First World War, its aftermath, and the rise of Hitler and Nazism, one would naturally be concerned about politicization or propaganda within any youth program, but Heisenberg seems to have remained blissfully apolitical. He certainly witnessed political and social unrest; his first opportunity to meet Einstein was thwarted by the killing of Einstein's friend and Einstein's quite reasonable absence. When Hitler came to power just as Heisenberg was awarded the Nobel Prize, Heisenberg left his international friends wondering as he steadfastly chose to stay in Germany, even as head of its nuclear program during the war. One of his best friends, Pauli, *was* a Jew and he himself was enough of a physicist to reject the Nazi/Aryan characterization of relativity as *Judenphysik* (Jewish physics), earning his own public denunciation as a "White Jew." There was never any hint of that race distinction in his relationship with Pauli. In modern America, he might have been a *wiggah*.

With his father on the faculty at the University of Munich, Heisenberg did have the opportunity to attend the home-town university with connections. Considering mathematics or physics, he was sent to visit Professor Lindemann in mathematics. He found an old man with a yappy dog on his desk and was told that he had ruined his mind for mathematics by reading Weyl, a well-known theoretical physicist. He was welcomed by Sommerfeld and quickly met Pauli, who in fact did sometimes go to classes. Pauli, ever respectful of Sommerfeld, did nevertheless compare him to a "Hussar colonel"; the military look, demeanor, and expectations often set Germans apart from other German-speaking people. Just as we Americans fail to distinguish Russians from other Soviets and English from other Brits, we have a tendency to lump all German-speaking people and sometimes more into *German*. Einstein as a Swabian Jew who twice denounced and renounced German citizenship is a puzzle to us.

Heisenberg had been encouraged to compete with an older brother, and had found his niche of mathematics and science to suit him well. It must have been hard to enter university and find some genius like Pauli to compete with. It set the bar high, but Pauli's knowledge, criticism, and confidence under the encouraging Sommerfeld must have also contributed to his education. When Heisenberg got a chance to hear Bohr, he was enraptured by the depth of Bohr's thoughts, but also prepared to challenge Bohr on topics he knew. This immediately gave an opportunity for the first of many long walks and talks and the opportunity to serve as Bohr's assistant, a prized position. He also landed a position at Göttingen with Max Born (1882–1970) before he finished his doctorate at München. Göttingen was the mecca of mathematics, but was probably second to Berlin in Germany's world of theoretical physics. Heisenberg said he learned optimism from Sommerfeld, mathematics from the people at Göttingen, and physics from Bohr. Add Pauli to that list as critic/friend and it is hard to imagine a better education at that time. His competitive spirit, hard work, and humbling experiences developed him for the early success which he achieved. Experimental and theoretical physics was not so clearly differentiated by that time, and Heisenberg was required to do a year of experimental physics with Wilhelm Wien. When he came to his final oral exam with Sommerfeld, Wien, and others, he excelled in theory earning an A from Sommerfeld. Wien's questions about the resolution of optical instruments stumped Heisenberg, yielding a solid F. (He actually received, according to the rules, only one grade, C, the presumed average of an A and an F.) His friend Pauli of course received an A in spite of his later reputation of even hexing all experimental equipment. Because of Pauli, it has become standard fare in physics that theoreticians cannot do experiments and should be kept away from any laboratory equipment. Heisenberg was mortified by his near failure, even offering Max Born the opportunity to rescind the job offer at Göttingen. The great and wonderful irony is that Heisenberg's best-known success, his uncertainty principle, was explained in terms of Wien's question, which he had failed. We will probably never know how much Western civilization's survival of the potential threat of what turned out in fact to be a paltry German atomic bomb program under Heisenberg was due to Heisenberg's ineptness with experimental physics, or Wilhelm's Wien's inability to ask probing questions on nuclear physics, or Pauli's ability to hex an experiment even at a distance. There are *many, many* other reasons for Germany's failure and the corresponding success in America, but I would like to believe that Heisenberg did not want to build the bomb and that he used his position to discourage if not sabotage the effort. It was hard for many, even those closest to Heisenberg, to give him the benefit of the doubt.

In the summer of 1925, almost exactly a quarter century after Planck's discovery, quantum physics was still very muddled, with lots of successes but certainly no coherent form, no complete theory. What kind of principle or conceptual basis could produce something to provide any unified account of the impingement of quantum steps into the continuous classical world and the failure of classical concepts at the atomic scale? We tend to forget or erase our failures. At the very least we do not put them in our textbooks or tell our students. Such is the case of a paper by Bohr, Kramers, and Slater. Not taught in our physics curriculum, this may have been the last gasp of classical physics, allowing or even enabling a fresh new gasp for the baby quantum physics. And the all-time quantum hero, Niels Bohr looks like the goat.

For the first time in our story, an *American* comes with a significant contribution: Arthur Compton, later to be awarded the Nobel Prize for his experimental work, the scattering of gamma rays from electrons. Though his work was repeated in Europe (by Debye) and would naturally have been credited there (especially since he published first), a well-known European physicist (Sommerfeld) had heard of Compton's work on a visit to America and had consistently and insistently given the prior credit to Compton. Recall that gamma rays are electromagnetic radiation of high frequency. When gamma rays strike matter (which has an abundance of electrons), electrons are knocked off at various angles, and the frequency of the gamma rays are down-shifted depending on the angle of scattering. This is recognizable as a high-energy version of Einstein's 1905 explanation of the photoelectric effect, except that the higher energy gamma-ray photons had more than enough energy to kick off electrons and bounced off instead of being fully absorbed. A high-energy photon, a gamma ray in the older language, strikes an electron, knocking it for a loop (well, in a straight line) and loses some energy as it bounces off. The old stand-bys from classical physics, conservation of energy and conservation of momentum, were sufficient to calculate the relationship of the angles of scattering and the frequency shifts, if we know the formulas from Einstein's understanding of photons. This is quite easy and is done by every second-year physics student. It is easy *if* you believe that electromagnetic radiation is in the form of particles.

Hendrik Kramers (1894–1952) was a Dutch assistant of Niels Bohr who surpassed Heisenberg linguistically and musically, at least quantitatively. He spoke several languages fluently and played the cello as well as the piano. We do not know his beliefs about photons, but he apparently saw and understood the relations of Compton scattering. There is some claim that this would have netted him a Nobel Prize. In any case, Bohr did not believe in photons and was willing to forego the seemingly sacred principle of energy conservation of energy in individual cases to argue for a weaker principle, a statistical *average* conservation of energy in the scattering of electromagnetic *waves.* Bohr won this argument (Did he ever lose one?) and published the paper with Kramers as a junior author. A new young arrival from America, J. C. Slater (1900–1976), also felt steamrollered as a junior author, but his concept of a "virtual oscillator" made it into the paper. The scandal going back to Bohr's "old quantum theory" was that the charged particles of any presumed classical motion did *not* oscillate with the frequencies of the electromagnetic radiation. Virtual reality did not originate with computerized video games; if you need vibrations at a given frequency, you can have virtual oscillators in the theoretical physics of the 1920s. The paper was roundly opposed; it did not ruin the reputation of any of the authors, but it was a solid failure, perhaps the death knell of classical physics. It failed the way all theories should fail, namely in the laboratory. These events of Compton scattering were sufficiently energetic to be seen *individually*, and energy conservation in *each* event was upheld. Compton's work is generally taught and believed to be the clincher for the particle nature of light which Einstein had so boldly stated twenty years earlier. It could be considered a baptism and christening for the photon, whether you believe in infant baptism or adult baptism. The naming did also come in 1925, but not of course by Bohr, Kramers, or Slater.

Heisenberg had all the tools: the optimism of youth and of Sommerfeld, the mathematics of Göttingen, the physics of Bohr, and the skeptical criticism of Pauli. The problem had been sticking in the craw of physicists for at least a decade: how can you combine h and its quantum jumps with the world of classical physics of continuously moving particles and oscillating electromagnetic fields which had worked so well? He also came with a principle he thought he had learned from Einstein's work, and also had the freedom of youth (and the example of Bohr) to throw away cherished principles even while holding theory's feet to the fire requiring "every detail [to] be subjected to the inexorable test of experiment." The principle he saw in Einstein's work was the positivistic lesson to deal only with what you can sense, what is objectively operationally observable. What could he give up? If we are to believe later recollections, it would seem that he may have been eager and excited to give up that third rail of classical physics, *causality.* Kant's supposed resuscitation of classical science from the ravages of Hume's skepticism had involved the CPR of the human mind's intervention and supply of *space, time,* and *causality.* Einstein had exorcised the space container, called ether, and warped time beyond recognition; why should he, Heisenberg, the next great mover and shaker, not exorcise causality? It would at least be *revolutionary.* Armed with the problem, the tools, and the attitude, all he needed for the revolution to appear as a revelation was to find an isle of exile.

The Isle of Patmos in the Aegean Sea, a small sterile island which became the center of the Orthodox church as the place where the Apostle John the Theologian wrote the apocalyptic Revelation, was not easily available to Heisenberg, but a smaller, more sterile island in the North Sea near Denmark was an ideal place to fight a very bad case of hay fever in the summer of 1925. The rocky island of Heligoland (literally "holy land," arguably) had been British in the nineteenth century but was ceded to the Germans in 1890. The treeless, therefore pollen-free, island was heavily bombed during the Second World War and, evacuated and remaining uninhabited from 1945 to 1952, was literally used by Britain as a bombing range, experiencing one of the biggest non-nuclear detonations in history. The ownership for centuries by Danish kings and the massive bombings could almost make one a believer in Jung's synchronicity; Heisenberg's connections with Niels Bohr and the Copenhagen hegemony in quantum physics and the German bomb project seem to make the additional connection, a hay-fever-induced-exilic-creation-of-quantum-mechanics on an island near Denmark, seem almost inevitable. (I seem to have caught the German talent or disease of prolixity in a single word, *Heuschnupfeninduziertvertriebenequantenmechanikschaffung*; for translation, see the multi-hyphened word above.)

After the hay fever subsided, the quantum fever still left him with at least one more sleepless night. The classic problem of atomic quantum physics had been the nearly simplest problem known, the presumed planetary problem of an electron orbiting a proton (the hydrogen atom) and the electromagnetic radiation connected to it. In true physicist style, Heisenberg had realized that problem was too difficult and made an attempt on a simpler problem. The ridiculously over-simplified idealized problem is the standing joke which is funny to physicists because we know how limited we are and how truthfully this is often our only MO. Faced with a difficult problem which in fact should require engineers, economists, or maybe even politicians and psychologists, physicists often make real progress by starting with two standards: "Assume a spherical cow (or any other object of interest)" and "Reduce it to a simple harmonic oscillator." If it takes up space, a sphere is a good starting shape; if it moves, it is a simple harmonic oscillator (SHO), or at worst a series of virtual oscillators. There is in fact a theorem introduced in the nineteenth century by Fourier which asserts the possibility of analysis and synthesis, i.e., breaking down and building up, of any time variation by these perfect sinusoidal oscillations of SHOs. (SHO is the ultimate shibboleth for a physicist; if they don't know SHO, they are not physicists.) Fourier's theorem gives credence to the physicists' claim that all problems reduce to one, the SHO.

Heisenberg knew that, practically speaking, we had no chance of seeing or measuring this presumed orbit in an atom and he knew, theoretically speaking, such a presumptive orbit led to nonsensical results such as collapsing atoms and meaningless frequencies. So he purposed to deal only with what we *did* or *could* know, the frequency and intensity (brightness) of the radiation (electromagnetic waves) coming from the quantum system. Since the oscillating fields did not have obvious and immediate connection with anything we could visualize, he suggested they were virtual oscillators which we call Fourier components used to make up the motion of electrons. In other words, he tried to get at the *inaccessible* motion of an electron by virtual oscillators, equally fake inaccessible pictures of what you get only as electromagnetic radiation. He could not do the atom problem and the SHO seemed too simple to get any important results, so he worked on the not-so-simple non-harmonic oscillator, the next level up from the simplest; we call it the *an*harmonic oscillator.

Heisenberg repeatedly renounced or gave up on visualizability (*Anschaulichkeit*) in his Nobel lecture of 1932: "renounce," "forego," "relinquish" the visual image of an electron path; he yielded to a *fundamentally unvisualizable microphysics*. He did this in spite of the experiments involving the discovery of electrons showing *paths* of electrons with cloud chambers, fluorescent screens, or discharge tubes, pictures which made electrons appear to act the same as "big" classical objects. Understanding usually means visualizability, i.e., seeing some image of what is going on. Here the birth of a quantum theory comes from Heisenberg's explicit (at least in retrospective reflection) claim that they "can be understood only by largely foregoing a visual description." He did use a weasel word "largely," admitting that he clearly did not *completely* forego the concept of the path of the electron. He may be said to have been in error to suggest that we must forego that concept, but it was an error which forced or allowed him to come at it from the back door, in effect tying the electron's path to the world of oscillators, virtual oscillators or Fourier components. He was basically denying any waves or particle or paths or *anything* visible for the electron in the atom, but he was *effectively* and almost certainly *unwittingly* sneaking waves into the world of particles. When you do this is theology, i.e., having insufficient language but forging ahead with what you know to talk about, it slips into vague theological language which seems, as philosophy often does, designed to cover over your ignorance and some very real mysteries which would make your ignorance totally reasonable. Heisenberg's language of renouncing visualization to gain understanding is reminiscent of the old theological conundrum of believing in order to understand, of seeing with faith's eye in order to get all the sight we have. Going to the limits of our physical eyes is only the beginning; we must go on to the invisible, to what we see by faith and by revelation. When we do this in physics, the reality or even the un-reality gets swallowed up in vague *mathematical* language. Well, if you are lucky and clever, you make very exact, very compact, very self-consistent mathematics which seems to stand on its own and dictate to the physical world. Thus many physicists will tell you that Nature is just exactly what the mathematics says it is. That is like theological terms becoming proscriptive for God rather than descriptive. The mathematics looks firm, solid, and self-sufficient, but we often do not *know* what it means.

As Heisenberg worked through the complicated mathematics with which he was trying to reconstruct the old familiar ideas of position, velocity, and energy, he eventually arrived at a strange form which worked out. Theoretical physics builds bridges out into thin air and sometimes they seem to land on something solid. When they do, you don't know whether you've hit pay-dirt or fool's gold. Do you go back to Sommerfeld, your encouraging first teacher who was also quite good with mathematics? Do you go to Born at Göttingen where the mathematicians were so good that they did not think they could trust the physicists with it? Or to Bohr who was the master of atomic physics, enslaved already for years in his quest for quantum sense in the atomic world? Or to your classmate/friend Pauli, who so consistently cut the legs out from under the ideas of lesser lights, i.e., everybody.

It happened that Pauli had moved to Hamburg. His late-night schedule, earlier dedicated to prodigious feats in physics, apparently developed

a broader range including sweets, alcohol, and women in Hamburg. He later said that in his youth (in reverse of his later years), physics was easy and women were hard; he was still quite young and the night-time mix of women and alcohol made him later remark that there was a complete disconnect between his day-time and night-time relationships with women. A night-life district in Hamburg named Sankt Pauli (you may know it from the beer named after it) was a natural source of an appropriately twisted nickname, even though it pre-dated Pauli's appearance. Pauli's reputation for hexing experiments also took on greater power in Hamburg where Otto Stern, the famous experimentalist, was a good friend who would not allow Pauli to set foot in the lab. Hamburg is close to the North Sea where Heisenberg had his hoped-for epiphany on the barren island. Thus it came to pass that Pauli was the first to see Heisenberg's work.

Contrary to some natural expectations, Pauli was encouraging and encouraged. He saw a ray of light and hope in Heisenberg's work, feeling he was now able to go to work. Heisenberg's paper was forwarded to Born, who was too busy to look at it immediately. When he finally spent some time with it, he recognized the mathematics as matrix algebra, something he had encountered decades earlier. The strange feature of non-commutativity of matrix multiplication, i.e., getting different answers multiplying in different orders, $AB \neq BA$, became a central feature of what came to be called matrix mechanics. Born, recognizing the mathematics and the importance of the formalism, wanted to develop the mathematical theory. He asked Pauli to help; Pauli declined, suggesting with a clear note of sarcasm that the Born/Göttingen mathematical development would obscure and lose the physics. Born found a young mathematician/physicist named Pascual Jordan (younger even than Heisenberg) to be the hired math gun and initiated a collaboration (mostly by correspondence since they could not get extended time together), producing a famous follow-up paper to Heisenberg's discovery, called the *Dreimännerarbeit*, the "Three-Man-Work."

Pauli, with a brilliant mathematical mind, was still distrusting of the emphasis on the mathematical formalism that the Göttingen combination of Born and Jordan brought to Heisenberg's work. Just as a musical performer has to supersede the musical training and get it into their gut to express themselves artistically, Pauli seemed to sense that Born and Jordan would be stuck in the formalism and fail to express the underlying physics. Mathematical genius is not enough; sometimes it may be necessary but it is never sufficient. Heisenberg wrote to Pauli on October 23, 1925 regarding the paper which Born, Jordan, and Heisenberg were writing: "[F]ormally it is still very beautiful. . . . The most important part of the paper seems to be, *in spite of* [his emphasis] its mathematical character, the principle-axes transformation. . . . [U]p to now it [the theory] is only of little help." He was feeling keenly the inability of the theory to deal with some most basic physical problems; specifically it had not even risen to the level of solving the hydrogen atom, the Balmer series in the spectrum, the old Bohr atom. Pauli, perhaps gloating that his assessment of Born's formalism spoiling the physics was correct, answered: "I could hug you for this *in spite of.*"

In spite of his seemingly unerring sense, Pauli had not contributed positively (except for encouragement). Heisenberg provoked him, calling him a "jackass who had not accomplished anything physically new." Pauli, who with great regularity had punctured the pride of others, was now wounded himself. Determined to beat the exalted Göttingen mathematicians at their own game or perhaps shut the game down as a failure, he aimed to pick "the pearl of physical wisdom out of the chaff of mathematical formalism." In less than three weeks, he solved the problem that Born, Jordan, and Heisenberg had failed to solve. Born, often feeling incompetence in the face of Pauli's brilliance, said: "It was stupid that we could not solve the hydrogen atom; it was straightforward." Yes, after Pauli showed the way. Not only had Pauli shown his own mathematical virtuosity, he had pulled the horribly abstract, formal, and complicated mathematical system from the abyss, rescuing the physics out of the rotting apples of mathematics, at least for himself, a mathematical genius. The derivation of the Balmer formula, formerly only a numerological patch on the simplest of the massive spectroscopic data in 1885 and the keynote success of a slap-dash hodge-podge classical atom with quantum patches in Bohr's 1913 work, was decisive. He had converted himself not just to the physical insight of Heisenberg, but also to the matrix formalism. He was thrilled by "*die Heisenbergsche Mechanik*" but had also found the "Göttingen deluge of learning" (*Göttinger Gelehrsamkeitschwall*) to be useful, at least in his own hands.

Paul Dirac, here as in real life, can easily be ignored. Raised with a demanding father who taught French in England, he was required to speak French correctly or not at all. He thus developed a habit of speaking almost not at all, in English or in French. A *dirac* was defined as one word per hour, his presumed rate. He did not mesh well with the voluble Bohr when he was chosen for a brief stint as Bohr's assistant. Bohr worked by speaking about what he did not understand. This resulted in one-sided conversations with many versions of unfinished sentences. Dirac said: "I was taught at school never to start a sentence without knowing the end of it." He was known to lecture by reading word for word from his book. After all, he had thought it through once and decided this was the best way to say it. When there was a complaint about Dirac's silence, the story was told of shopping for a parrot. The first parrot had a reasonable vocabulary, but seemed expensive. The second parrot spoke fluently in several languages, and the price was clearly out of range. The third parrot was scruffy and didn't speak at all. Maybe that one is affordable. How much? $100,000. What?! Oh, but that one thinks. It is wonderful that "an Edwardian geek" who doesn't speak can find a place in the physics world; he can if he thinks like Paul Dirac.

Paul Dirac, also younger than Heisenberg, being in the boonies of Cambridge University, managed to get an advance copy of Heisenberg's paper. After a few weeks of not making sense of it, he quickly developed a formalism and a notation for the new mechanics. He even managed to publish on it before Heisenberg's paper came out. When a few years later, both Heisenberg and Dirac wrote books explaining this new QM, Heisenberg's book faded into obscurity while no physicist's collection is complete without Dirac's book.

I have never seen Heisenberg's book, but every physicist must read Dirac. It is remarkable that Pauli saw through the maze of mathematics to a physics which could hold hope for him. Matrices have been very fruitful in systems with only a few possible states, basically ladders with a few rungs, more like kitchen foot-stools. They have not been easily decipherable for systems with essentially an infinite number of infinitesimally close rungs which position and velocity (or momentum) of free particles approximate. Born was apparently the first to write the formula for the commutator ($AB—BA$) of position and momentum, as the quantitative measure of *non*-commutation; later this came to be seen as the source of the uncertainty principle, a core of QM. Born, touting this as one of his major contributions, felt slighted as he waited two decades before being awarded a Nobel Prize of his own.

Just as the invisible world of the spirit seemed to naturally produce splits between the East and the West, between Catholicism and Protestantism, leading to literally thousands of splits, the invisible atomic quantum world seemed destined to have more than one way to go. Within the year of Heisenberg's theory of 1925, a completely different approach had similar if not greater success. This is our next topic, wave mechanics.

10

Another Form of QM and the Loss of Determinism

> The waves of light and the particles of matter become the wave/particles of light/matter, cementing the indeterminism of our theories of Nature.

WHAT DID THE OLD folks think of this new formal theory of QM? Planck at sixty-seven and Einstein at forty-seven were in Berlin, which still laid claim to being the chief citadel of physics in Germany. They invited Heisenberg to speak, and Heisenberg and Einstein had their first extended conversation. Heisenberg insisted that he was following Einstein's strategy, building on what one can actually observe, not on unknown or even unknowable quantities or concepts. Einstein wasn't buying it. Observers observed real physical events in relativity; there was an objective existence. Heisenberg was acting as though the structure and behavior of real structural elements of the atom were meaningless categories. The abstract, even obscure, mathematics of matrices did not illuminate anything for Einstein. The traditional questions of "Where?" and "When?" were still legitimate questions for Einstein. Heisenberg was willing to let those questions go and wait for new questions and concepts to take their place. Young but confident by now, Heisenberg held his own but neither would be convinced.

A new strand appeared from another little-known village of Europe, Paris. Not to suggest that France did not make contributions to theoretical physics—Poincaré (1854–1912) is a great counter-example at the turn of the century—but it appeared somewhat in the backwash of this era dominated by the Germanic tribes. Can any good theoretical physics come out of Paris? Paris may have rescued a young Polish girl named Maria Salomea Skłodowska (1867–1934) and made her into an experimentalist *par excellence* (known to us as Marie Curie), but could it produce a theoretician of note from a tradition of statesmen, politicians and military officers? Maurice de Broglie (1875–1960), of noble birth and the son of a member of parliament, came of age when Becquerel's career and the world of physics rocketed into a new era with his nearly accidental discovery of radioactivity. Although Maurice's accomplishments are not to be sneezed at, he was fated to be remembered primarily as "the older brother."

Becquerel (1852–1908) had comfortably settled into his position of chair of physics at the Musée d'Histoire Naturelle, following his father and grandfather. Already in his forties and balding, he had settled into some bread-and-butter physics of phosphorescent minerals, an unlikely launch pad for a career move. When Poincaré received a letter on January 7, 1896 containing X-ray images of Röntgen's hand, the whole world got excited. Only three weeks later, a young Eddie McCarthy of Dartmouth, New Hampshire made international news by having a broken arm set with the aid of X-ray images. By January 20, two local doctors of Paris had produced X-ray photographs and shown them to the French Academie des Sciences. Becquerel, seeing the ghostly images, tried his phosphorescent minerals as sources of X-rays and accidentally discovered radioactivity. With the Curies, he received the newly established Nobel Prize only seven years later. Perhaps as important for the course of physics, the flurry of activity in this new field of radioactivity attracted the young Maurice de Broglie. Against the tradition and wishes of his father and grandfather, he turned to the world of science while his kid brother was barely out of diapers.

Born in 1892, Prince Louis de Broglie (little brother to Maurice, who somehow is not usually referred to in those grand titles that Louis inherited only in 1960 on Maurice's death) studied history at the Sorbonne, preparing to become a diplomat. His brother's influence and a practical stint with electromagnetic practice in a mobile radiotelegraphy unit during the war headed him belatedly toward a career in physics. Perhaps his position outside the mainstream of the quantum world compensated for his lack of youth by the time he did a doctoral thesis. In any case, the not-so-young thirty-one-year-old in 1923 brought a fresh idea.

He knew that Einstein had suggested that light could act like a stream of particles. Would it not be possible that particles, those pieces of matter that we can easily *envision* as particles, could act like a wave? Taking the Einstein-Planck equations which gave momentum and energy of the purported particle of light from its wavelength and frequency, he turned them on their head suggesting that the wave features of particles, wavelength and frequency, could be calculated from the momentum and energy. Matter and non-matter took on a symmetry which is still difficult for us to fathom; particles and matter could take on wave properties. The algebra was simple enough for a schoolboy; the symmetry was attractive. But otherwise, why should we care for such mythological speculation?

De Broglie, almost a full decade older than Heisenberg, Jordan, and Dirac and a mathematical minor-leaguer, planted a seed in two papers near the end of 1923. No soil, no watering, no germination. A year later, his thesis committee passed him, but were left scratching their heads about the significance. One of the committee sent a copy to Einstein, who was encouraging, a major step even if Einstein was not at the heart of the young movement. Being over-thirty and encouraged by an over-forty like Einstein did not appear to be the road to fame. Physics yawned. Enter an older thirty-something, of affluent if not noble birth, the Viennese Erwin Schrödinger (1887–1961). He too had been involved in the war, winning a medal. Entering the University of Vienna just after Boltzmann's suicide, his own professor had also died prematurely, in battle, leaving Vienna somewhat unattractive in physics for the child prodigy, Wolfgang Pauli, whose father was on the faculty there.

Raised by women, his mother and her two sisters, as an only child, Schrödinger found a wife in 1920 who stayed with him in spite of his string of women before and after. The marriage was childless, but Erwin (pictured below) found time to father three children with three different women and to father a theory of QM while spending Christmas holiday of 1925 at a resort with one of his girlfriends. The name of the girlfriend is triply lost: lost in the multiplicity of girlfriends, lost by code-names used in Schrödinger's papers, and finally lost because his private papers of that time were destroyed. Having developed considerable mathematical skills and a body of well-regarded theoretical research in a wide range of topics, Schrödinger had settled into a comfortable position in Zürich. Although he was not a classicist, he had found the trends in atomic physics distasteful. The discontinuity of Bohr's

jumping electrons bypassed all ideas of causation, and the orbital frequencies of the electrons had no meaningful connection with the frequencies of electromagnetic radiation involved. As a specific example, the lowest energy level of hydrogen in Bohr's model has an orbital frequency of 6.59×10^{15} Hz (i.e., cycles/second), the second level gives 0.82×10^{15} Hz, but the transition produces light at a frequency of 2.47×10^{15} Hz. This is totally unsatisfying; for waves of frequency 2.47×10^{15} Hz to be produced, something has to wiggle at that frequency. Heisenberg might casually insert a virtual oscillator and make a complicated mathematics which shows some consistency, but the lack of physics and the presence of ugly mathematics were repugnant.

Schrödinger became aware of de Broglie's waves. They were not to be cavalierly dismissed because they offered a real beginning for understanding the otherwise distasteful Bohr atom. Thinking of orbits as standing waves in a circle, the wavelengths of de Broglie matched the circumference. An integral number of wavelengths gave specified possible stationary orbits, specified by Bohr's artificial condition. (See picture below.) Moreover, although the frequencies of the "electron wave" as calculated by de Broglie did not give the orbital frequency; the *difference* in the de Broglie frequencies of two orbits gave a correct electromagnetic frequency, suggesting a beating phenomenon between two frequencies. While Heisenberg was curing his hay fever on the island of Heligoland, Schrödinger wrote a paper about de Broglie's electron waves, suggesting that the electron was a "whitecap" on a wavy sea of electron waves. Perhaps it could be thought of as a literally superficial manifestation on the top of continuous fields; we might avoid the distasteful discontinuity introduced by particles.

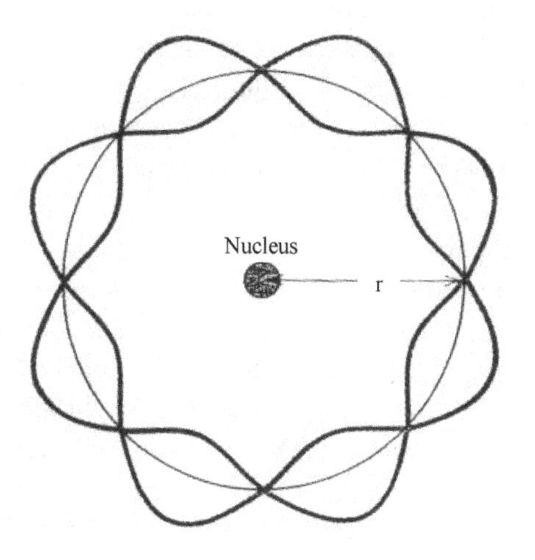

Schematic picture of four de Broglie wavelengths in one orbit of an electron about a nucleus.

A colleague (Peter Debye (1884–1966), who made a name for himself as well, winning the Nobel Prize in chemistry in 1936) posed what should be the obvious challenge: write the wave equation, a differential equation, governing this wave, just as all classical wave motions had such a governing equation. (In spite of this common story, Debye refused credit for this contribution, saying only that he pushed Schrödinger to give a seminar on de Broglie's paper.) This is when Schrödinger, in the "late erotic outburst" of note, took his girlfriend and his idea on a holiday break from the university and his wife. He came back with all the workings of an alternative quantum theory. In rapid succession in 1926, he published a series of papers which are recognizable as a large portion of the present-day standard course in QM.

Schrödinger produced a reasonable wave equation by guesses from classical physics. Applying the theory to an atom with reasonable conditions, he found wave patterns which corresponded to discrete energy values. Although more complicated, they were as natural as the simple set of vibrations for a stringed instrument, the original basis for music theory set down by Pythagoras thousands of years ago. Of course, they match the values of the successful Bohr atom, giving a basis for the Balmer series among others. Moreover Schrödinger could see great possibilities of a smooth transformation of one pattern into another, avoiding the quantum jump which he detested.

The really old timers, Einstein at forty-seven and Planck at sixty-eight, welcomed Schrödinger to Berlin for a visit. Einstein was enthusiastic about the advance of de Broglie's waves, which he had already described as the beginning of the lifting of the fog. There was a chance of classical concepts being restored. There was both a generation gap and a philosophical gap between Heisenberg and Einstein. Einstein was less than a decade older than Schrödinger; they had some similarities in their bohemian life styles, to try to put it delicately; most importantly, they shared much in their concepts of physics, including the distaste for aspects of the Bohr-Heisenberg picture which was on the rise. The Berliners' respect for Schrödinger was manifest soon thereafter by his appointment to the chair of theoretical physics replacing the venerable Planck, who retired.

One year before, in 1925, we had no true quantum theory. Now, by early 1926, we had two competing theories. No matter that Pauli and Schrödinger quickly proved (by April and May 1926) that the mathematics which looked so different were formally the same. Both Heisenberg and Schrödinger thought they were fighting for the soul of science: Schrödinger for the soul of the old system; Heisenberg for the new system. Schrödinger promoted real waves while Heisenberg wanted to get rid of waves completely in the microworld. Heisenberg had pushed the observational elements to the fore and pushed the undetectable electron motion to the background; Schrödinger made an underlying undetectable wave to be the frontispiece of his work, symbolized by ψ, which became the enduring symbol of quantum physics.

Heisenberg's reception had been tentative. After all, Max Born may have been the only physicist alive who knew matrices, and the arcane mathematics did not illuminate the physics nor did it yield results easily. Schrödinger's reception was overwhelming. Every physicist had studied waves; every physicist had learned differential equations. Even Max Born, Heisenberg's mentor and collaborator, exulted that he was ready to "defect—or, better, return" to continuous physics, to "the crisp, clear conceptual formulations of classical physics." Heisenberg himself had to concede that the mathematics was easier; in his calculations, he gave succor to the enemy by using Schrödinger's approach to get past calculational difficulties.

The controversy brought out strong language. Schrödinger had made no bones about the disgusting, repugnant, and repulsive nature of matrix mechanics. Heisenberg called the wave business "crap." Schrödinger, perhaps because of confidence of the wave at his back, appeared gracious; Heisenberg appeared desperate as though it were a personal attack. Schrödinger expressed hope that wave mechanics and matrix mechanics "will not fight against one another, but . . . will supplement one another, and that one will make progress where the other fails."

The calculational and visualization success of the wave mechanics was all too sharp against the failures of matrix mechanics, even touted for its avoidance of visualizability. Heisenberg, according to his foes, had laid a large quantum egg. He must have seen his newfound fame quickly melting away.

When a large conference was held in Munich in summer 1926, Heisenberg returned to his alma mater to visit family *and* to confront Schrödinger. When Schrödinger spoke, Heisenberg rose to point out that this wave conception could not even explain the Planck-Einstein relation of energy and frequency. This was Heisenberg's hometown, but friend and foe alike turned against him. Wien, the experimental physicist who had tried to flunk Heisenberg on his dissertation, got another shot in, suggesting that Schrödinger would soon clear up any difficulties. Heisenberg: "My arguments had clearly failed to impress anyone—even Sommerfeld, who felt most kindly toward me, succumbed to the persuasive force of Schrödinger's mathematics." And Heisenberg's arguments against the waves were proved wrong as the wave theory continued to succeed.

Warning Bohr that the quantum empire built since his atomic model of 1913 was collapsing, Heisenberg retreated to Copenhagen. But it was no picnic with the dogged Bohr, even though there was some basic agreement. Heisenberg was fighting the waves furiously while Bohr was struggling with the bigger wave/particle questions. Inviting Schrödinger to Copenhagen in October 1926, Bohr pursued his points "with a fanatic, terrifying relentlessness" (Heisenberg's words). Bohr was relentless even when Schrödinger fell ill and was confined to bed. Schrödinger exploded: "If you have to have those damn quantum jumps then I wish I'd never started working on atomic theory!" Galileo may not have had it nearly that bad in spite of the stories we hear; he only had to deal with his "friend," the Pope of the Catholic Church, but Schrödinger was sick in the Vatican of Quantum Theory with Pope Bohr bearing or Bohr-ing down on him. This was not just a battle of personalities about questions of priority or fame. There was a sense that the shape of Nature, the shape of physical law, the shape of the Universe was at stake. The philosophical ramifications loomed large in their minds and we are still struggling with them, large or small.

Schrödinger, by the calculational success and by supplying even the weakest reality for the virtual oscillators of a Bohr-Heisenberg atom, was clearly winning the battle. In a sense, he embodies the long-term success of QM based on pragmatic success, even when the deep understanding is minimal. But the Copenhagen hegemony was winning the war. Schrödinger may have built the best calculational machine, but Bohr and Heisenberg were able to jump in the driver's seat.

The success of the wave mechanics brought the seeds of its downfall. Having *something* to wiggle at the right frequency seemed like a big step forward, but eventually we have to ask: *what* is wiggling? The answer is simple: ψ, the symbol for Schrödinger's wave function. ψ is the Greek letter *psi*, usually pronounced "sye" in America while I, to the ridicule of my students, have persisted in sounding the *p*, as the Germans do. ψ is arguably *the* icon of QM, and appropriately so, but no one understands QM and no one knows what is wiggling. Giving it a Greek name does not give it reality. If Einstein and Schrödinger had known what would become of ψ, they would have been singing a different tune: the large quantum egg was not laid by Heisenberg; it was ψ itself. Electric and magnetic fields were capable of an elopement and marriage, running off into space away from their sources, the charges; ψ was the prodigal son who never returned to his loving father Schrödinger.

Three contributions sewed up the meaning of QM in 1926–27 and left the old men, de Broglie, Einstein, and Schrödinger, alienated from "orthodoxy" for life: Max Born pronounced the meaning of ψ, Niels Bohr propounded a principle of complementarity, and Werner Heisenberg enunciated the uncertainty principle. This was the three-legged stool on which the Copenhagen interpretation rests.

Schrödinger had tried hard to make ψ real. ψ's wiggles were the *real* oscillations in the atom to displace the vaunted virtual oscillators which proudly carried the right frequency and intensity, even if they otherwise were totally phony, *or* ψ represented deep waves whose surface white caps were the particle manifestations *or* ψ was a wave packet, a wave squeezed into a small space-time to mimic a particle. For various reasons, none of these conceptions succeeded. The hoped-for reality succumbed to a much less vaunted place, a very imperfect knowledge. In the language of philosophy, the two schools of thought talked past one another for decades, one speaking of ontology and the other speaking of epistemology. One group was going for understanding of the reality behind it all; the other was going for a description of what we can know. They were doomed never to hear one another. ψ was interpreted as a probability amplitude, whose square gave the probability of a position or of one of the choices being considered in an observation (Born's interpretation). It raised imperfect knowledge, probabilistic laws, to a fundamental status and perpetuated the jumping business that Schrödinger so disliked. All we can know, even if we know them perfectly, are the betting odds. ψ, cast in the classical mold of differential equations, could be perfectly known and would evolve deterministically, but all it gave were the *possible* values of physical variables and the exact *probabilities* for each. When the race is run, the Las Vegas betting odds are out of date and a result is certain. A probability distribution (e.g., 1/6 for each face of a die) collapses to certainty (probability 1) for one value and no chance (probability 0) for all others when the die is cast. We have generalized the unexplained jump in Bohr's atom of 1913. The clever internal workings in the quantum world may not be free will but they are stubbornly unknown to us. That in itself is a big blow to the classical physics tradition of knowing it all perfectly. But there is an important extension: it is *unknowable* to us; it is unknowable *even to God*. That is what is called pure chance or unavoidable chance, whether we are correct or not.

Born later said that statistical considerations were common, and shifting it one layer deeper was not a big deal. His own words in his 1927 paper say otherwise: "here the whole problem of determinism arises. . . . [In] QM there exists no quantity which in an individual case determines the result of a collision. . . . I myself am inclined to give up determinism in the atomic world."

Loss of determinism is a big deal.

11

What Is This Thing Called "QM"?

Double realities form the core for both QM and Christianity.

WHEN I FIRST TOLD a group of my friends what I was writing about, the obvious question arose: what is QM? As much as I knew, having taught courses about it numerous times, I was stunned by my lack of a reasonably articulate concise answer. And I continue to be stunned by what I know runs much deeper than writer's block. One can run through the history of the development of ideas, one can recite the catechistic postulates or rules or even reproduce large segments of standard problems like serious Christians quote memory verses, but some of us know there is a vein of mystery that runs deep and a Sunday-School recitation fed to children as though it is *simple gospel truth* reducible to bumper stickers just cannot easily pop out of our mouths. I am always happy to speak of my faith in God and/or QM, but I am deeply aware that any smatterings of truth that may come from brief conversations or visible symbols easily get admixtures of misunderstanding. The darkness of the unseen unknown, our apparent need to see and understand within that void, and the desire to communicate and use the scraps of knowledge we think we attain even when they sound foolish or over-simplified, all this is so similar in our looks at the physical world (science) and our looks at the spiritual world (religion). If someone clearly knew nothing of my faith and asked sincerely: "what is your Christian faith?" I would probably be the same tongue-tied stammerer. Let's try a brief answer: *God is an eternal spirit who created a lawful world including agents free to break the law. They did. He then played double-agent coming into the world as a man, Jesus, to help us and the remainder of creation have the right relationship with him.* Not exactly twenty-five words or less nor does it come close to covering everything. But a few key words cover most of it: freedom (his and ours), double-agency (God/man), creation's bad choices (sin), and God's good choice (grace).

OK, let's try QM. *QM is the science of the ordered activity of a free physical Nature which exhibits the double-agency of discrete discontinuous systems (like particles) and continuous systems (like waves), a double-agency of necessity and chance.* It is tempting to say that QM is the science of motion (mechanics) at the atomic level. But that is like saying Christianity is what goes on inside a Christian church. QM in some of its manifestations appears necessary to make sense of the atom but it is not confined to the atom. We need, deeply desire, something which is universal, maybe absolute in some ways, a faith and a science that transcends our buildings and our laboratory specimens. It is strange that we want and believe our science to be exclusively the way and the truth, but moral law and religious belief can be totally relative and confined to our church, synagogue, mosque, ashram, sweat-lodge, or even worse, inside our individual mind or soul. Christianity is not a private religion between a person and God any more that QM is a law between Nature and each single constituent of the physical world. We are all in this together and the mystical nature of our being, of our relationships within the whole, and relationship to God will not yield to simple formulas. QM underlines the limits of our knowledge, the mystery of being, and the necessity of wholeness and relationship.

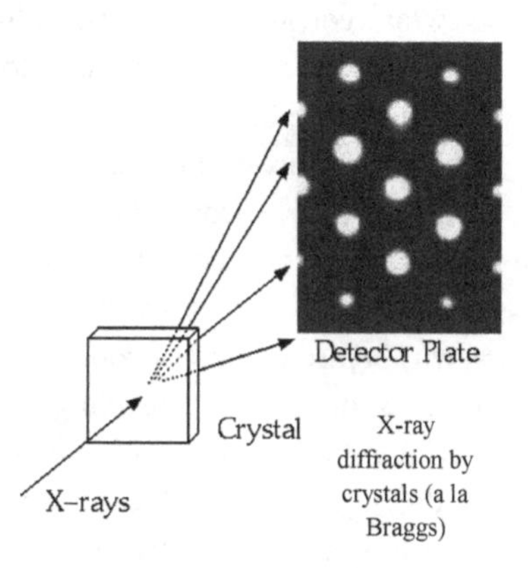

X-ray diffraction by crystals (a la Braggs)

* * *

When de Broglie posited the wave nature of the material world, he puzzled his thesis committee and his reviewers. Einstein had foreseen this fused wave/particle nature for *light* already in 1909 and was immediately positive, making some good use of the extension into the material world (Bose-Einstein statistics and condensation). More importantly, Nature confirmed the wave nature of the electron by interference experiments echoing the overturn of

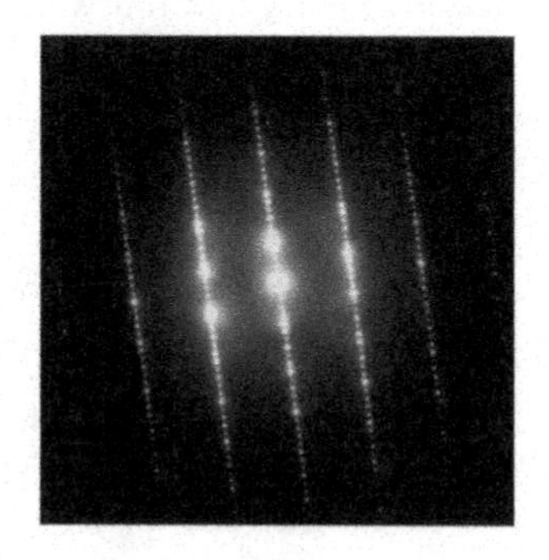
Electron diffraction by crystals (a la G.P. Thomson)

Newton's corpuscular theory of light by the experiments of Young and Fresnel in the early 1800s. Davisson and Germer in the US and G. P. Thomson in England did experiments with electrons scattering from crystals, repeating the X-ray experiments of the Braggs (father and son) a decade earlier. The father-son Braggs, Sir William Henry Bragg (1862–1942) and Sir William Lawrence Bragg (1890–1971), could collaborate and get the same result: X-rays behaved as electromagnetic waves of very short wavelength (about 1,000 times shorter than visible light). They thus shared the Nobel Prize of 1915. Father J. J. Thomson (1856–1940) had discovered the *particle* that we call an electron; son G. P. Thomson (1892–1975) discovered the *wave* which we call an electron. Both won the Nobel Prize (father in 1906, son in 1937) and the latter work confirmed de Broglie's work, which also earned him the Nobel Prize (1929), before the Nobel Prizes went to Heisenberg (1932), Schrödinger (1933), and Dirac (1933) for the theoretical application. As mysterious as such a dual nature of the physical world has remained, it may still be the best visualizable entrée into quantum nature. The God/man Jesus holds the same place of mysterious revelation in the world of Christian theology.

Just as Young's two-slit interference experiments with light countered Newton's corpuscular model, budded into the Faraday-Maxwell electromagnetic wave theory of light, and flowered into a dual nature in the quantum theory, the de Broglie-Thomson wave nature of electrons lifted the curtain on a dual nature of the material world. Schrödinger wrote and solved a wave equation for electrons. You would think we might be finished: a complete wave theory and a complete particle theory for both constituents of the classical universe, particles and fields. What more do you want? The unseen still yawns before us, and of course we cannot be satisfied. What *is* an electron? What *is* the wave of Schrödinger's wave equation? What is *waving* and what *medium* is it waving in? All are questions which rank up there with: Who is God? We cannot get off easily and quickly.

It is said that the two-slit interference experiments, of light or material particles, contain and illustrate all the mysteries of QM. Let's look at the two-slit *wave* picture and see why it feels incomplete or even wrong for an electron. If a wave strikes two slits and comes out the other side, it forms a pattern of alternately bright and dim spots (loud and soft spots for sound, big and small amplitude waves in general). (See Chapter 4 for a picture.) This is easily visualized by the following picture. We can even do very simple calculations to give positions and sizes of the large amplitudes and the small amplitudes. Waves are traveling fields which alternate above and below the ambient level, "positive" and "negative," usually called a "crest" and "trough" of the wave. When *two* waves arrive at the same place, they may have matching crests and troughs and thus add to a double-sized wave. Or they may have the crest of one matching the trough of the other and add a positive to a negative to give zero. The former case has waves *in phase* and produces

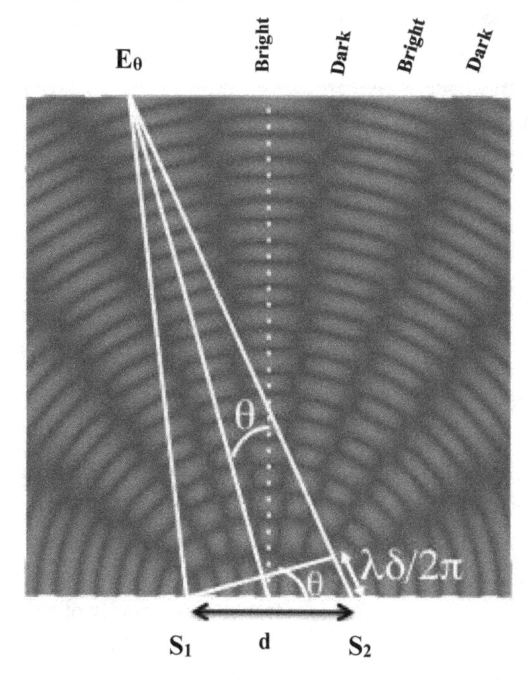

Light impinges (from bottom) onto two slits S_1 and S_2 separated by a distance d

what we call *constructive interference*; the latter case has *out of phase* waves and produces *destructive interference*, basically cancelling each other. And is it not clear that everything in between occurs, i.e., there is a continuum of phase relationships? Imagining a wavelength of one inch (for convenience), lay two rulers with their zeros at the respective slits (like tile corners on your bathroom floor) and measure to a point in space. If the difference is an integral number of inches, the two waves starting together from the slits will be in phase and will constructively interfere. This is effectively what the waves of the earlier picture shows.

What do we mean that electrons behave like waves or that this wave theory (the copy of light or sound wave theory) describes the action of electrons? We mean quite simply that if you can identify a wavelength associated with an electron, then there is agreement both qualitatively and quantitatively of this interference wave pattern and the statistical result of a *large number of electrons*. We, by the wave picture, get the *pattern seen in Nature*. You can see that we are resorting to the same idea that thermodynamics eventually lapsed into, statistics. When we send *one* electron, we don't know which slit it will go through or where it will end up on the screen. We know its statistical results, as though each and every electron knows how all other electrons will go or each one has some built-in propensities to tell it what to do. But try as we might, we cannot predict what each one will do. Why do we still call it a particle then? Why not just let it be a wave? Because we *see* it only as a *particle* when we *send* it and when we *receive* it, processes of emission and reception. In between, we are running on faith; it started as a particle and ended as a particle; we don't know what else to call it. This means that the particle of our faith does two strange things: it appears to go two (or more) paths at once and the ideas of probability are different from any particle we can conceive. Waves spread out and go all over and they add amplitudes. A classical particle is in one place at one time, one place has only one particle at a time, and the numbers (or probabilities) are what add and they end up being proportional to the *square* of the amplitude of the wave. Waves have phase and amplitude; waves simply add; the power or energy and thus the number of particles in a wave are related to the *square* of the amplitude. Point-by-point addition is just natural for waves. This simple addition we call superposition. A wave of amplitude 1 adds to a wave of amplitude 1 to give a wave of amplitude 2 *or* 0 *or* anything in between, depending on phase. *One* particle and *one* particle add simply to *two* particles. Particles (and all other quantum states that have this here and now, separable, countable quality) have to take on this strange new quality of the waves. A wave (or a wave function or state function, as we variously call them) must represent a statistical collection which adds 1 + 1 to get 4 (the square of 2) or 0 (the square of 0) or everything in between and of course averages to 2. We have to give up a treasured piece of logic, the law of the excluded middle. We believe a classical particle, any particle worthy of the name, goes through slit 1 *or* slit 2; there is no other way. But a thoroughgoing wave/particle duality, as we have in quantum physics, says that it goes through both and the interaction of the two "parts" involves amplitude and phase, wave-like qualities. If you cover one slit, you get a simple pattern; if you cover the other slit, you get a similar simple pattern, just scooted over. If you allow the particle to go through both, you do *not* get the sum of the patterns; the two patterns have a complicated interference pattern of the interacting square-roots.

* * *

Imagine you enter a room with other people. Each person is given a token, tossed onto the table. Each token looks like this and you are told that it represents one chance in a lottery for a door prize. Then they return giving each person an opportunity to receive another token, again tossed onto your table if you choose to have one. Then a third is offered as well. Below is a sampling of several people's tokens. They prepare the

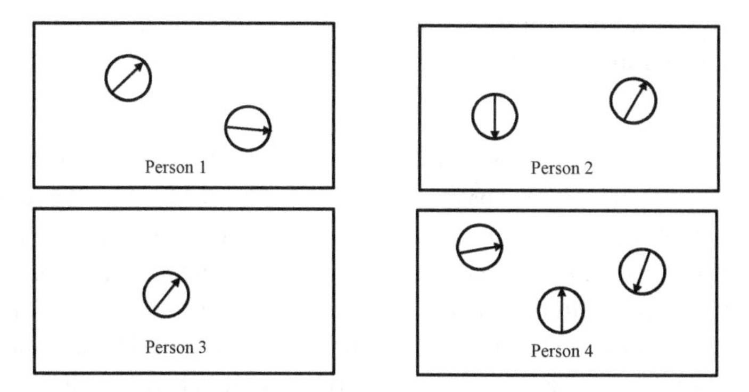

lottery, and looking at the tokens, they assign 2.56 chances to person 1, 0.16 chance to person 2, 1 chance to person 3, 0.36 chance to person 4. An immediate uproar breaks out. Person 2 yells: "Why do I get so little chance? I have two tokens. Why isn't 1 + 1 = 2? Why does person 3 get a full 1 chance and we (person 4 and I) get less chance with more tokens?" So they show how the chances are calculated.

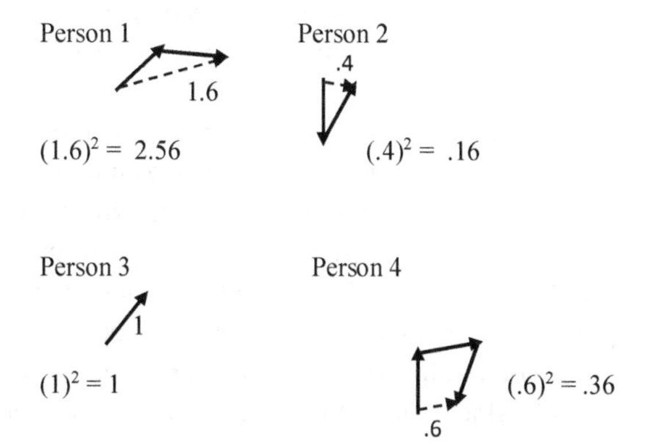

Why do you add them that way and why do you square them? Well, that is the mystery of QM! Because individual chances, acting like particles, do not add that way but waves with amplitudes (lengths) and phases (different directions) *do* add that way when you look at energy or the number of particles. A single slit gives a broad range of places having about equal chances for a particle; adding another slit gives extra chance to some places and takes away chances of other places. You can understand Bohr's willingness to horse-trade on energy conservation; some will gain and some will lose but it all comes out in the wash. QM allows probabilities to add in a way that we don't normally experience, adding like positives *and* negatives, more generally adding like vectors (our directional arrows) and being squared to always come out positive (or at least 0). And this way of adding probabilities seeps over into the realities, giving not only a richer game of probabilities but a richer range of reality.

* * *

To review, the key ideas of QM are 1) the dual nature of particles and waves and 2) the ability to have superposed realities with wave-like addition with amplitudes and phases. The thing which we call a particle is to us an indivisible individual entity which does not come in fractions; it is separable and countable, as an integral number. And yet, allowed to be itself without the typical questioning of its particle nature (such as "Which slit did you go through?"), it does not act like a particle at all but travels all the paths like oscillating fields in space and time. Not only is it infinitely divisible but the waves can coexist in space and time, exactly opposite of our puny idea of particles which take up *exclusive* space, not allowing another particle in the same location. The waves add, point by point in space and time, just as particles add. One particle here and one particle here results in two particles here, where we relax *here* to be big enough for several particles. But our best interpretation of ψ, the wave, is that its *square* gives the probability of a particle. That is, this interpretation agrees with all experiments we can do. The mismatch between the addition of the wave value (as all simple waves require) and the addition of the value *squared*, as the particle picture requires is a reflection of the simply irreconcilable picture of waves and particles; they are simply quite different beasts. Things which are created and destroyed as indivisible entities such as particles travel around and interact as waves. Waves that coexist in all space and time, adding by local values, carry energy and momentum and interact in physical processes like discrete chunks which we would call particles. It is apparent that we have not managed to conceive of a single conceptual being which is sufficiently clever for Nature's workings.

The writers of creeds from early church councils were crazy: Jesus was *fully* man and *fully* God?! But no crazier than those who say an electron is *fully* particle and *fully* wave. And maybe they are all crazy enough to be right.

Schrödinger, following de Broglie's lead, invented wave mechanics, trying for all he was worth to invest the wave with reality, but ψ, his wave function, could only muster the reality of a phased square-root of a probability. If you want the probability, square the amplitude; if two (or more) contributions are involved, add them with phase considerations before you square the amplitude of ψ. ψ may not be real but it is *determined* by the old classical games of calculus, differential equations. We maintain determinism but give up the reality. We know exactly what ψ is mathematically but we don't know what kind of physical reality it might have. Heisenberg tried to hang on to the measurable physical reality, the whole measurable physical reality, and nothing but the measurable physical reality. His end point was *matrix mechanics*. The mathematics looks totally different from Schrödinger's but it ends with all the same features, exactly. In either case, the formal theory allows us to calculate ψ, a state function or state vector, see how it develops in time, calculate the possible values of all possible measurements, and the probability of obtaining each possible answer. Classical mechanics allowed us to follow the development of all measurable quantities with *certainty* and with *infinite precision*. QM, the world of ψ, goes to a land of freedom with calculable limits and distribution shapes. It's a complete calculation machine for all the odds.

When the number of possible values of a measurement is uncountably infinite, an apparent continuum such as all points on a screen, the differential equations of calculus, Schrödinger's wave equation is the natural way to go; Heisenberg's matrices become nearly impossible. When the number of possible answers is small, matrix operators are not too ungainly, being simple arrays with the number of possible answers giving the number of rows and columns. For example, a system with three possible answers might use a numerical matrix like figure 1. The state function or state vector, called ψ in wave mechanics, will then be like a vector in 3-D space written as either a row or column of 3 numbers. For example, ψ might be figure 2 or (0 0.6 0.8). Quite contrary to Heisenberg's apparent intentions of sticking to the concrete measurable quantities, the resulting matrix mechanics seems to us more abstract with a "state" space (called Hilbert space) having the number of dimensions given by the number of possible measurement results and a vector representing the state of the system. The important idea is that the superposition of states, adding with amplitude and phase, which wave-like phenomena introduced to our particle world, carries over into this very simple but abstract world of vectors. Just as a particle could go through two slits, a very simple system with only a few possible answers can be in a state with *all* the answers, a state in which the answer is not specified. The possible answers are known, the odds are known, but nobody knows which answer will come up next.

$$\begin{bmatrix} 1 & 0 & 2 \\ 0 & 1 & -1 \\ 2 & -1 & 0 \end{bmatrix}$$

Figure 1.

$$\begin{pmatrix} 0 \\ 0.6 \\ 0.8 \end{pmatrix}$$

Figure 2.

Suppose we have two possible simple results, call them A or B (like "goes through slit 1" or "goes through slit 2" or "passes identity test A" or "passes identity test B"). By extending to any number of results, this may describe a quantum system with any number of possible results. We may represent a pure state A (certainty to produce result A) as $\binom{1}{0}$ and pure state B as $\binom{0}{1}$ as though they are forming a two-dimensional A-B space (see figure

3). The state of a system may in general be written as $\binom{a}{b}$ with $a^2 + b^2 = 1$. This is how the concepts of superposition and probabilities are expressed. It is now not *in* state A or *in* state B but some combination which will produce a distribution with a fraction giving A and a fraction giving B if we repeat a large number of times. A single measurement is never predictable (unless of course $a = 1$, $b = 0$ or $a = 0$, $b = 1$. This whole scheme of course seems almost too trivial to be of any interest, as we view it in our usual realistic frame of mind. You simply have two results A and B and you can know the odds for any measurement.

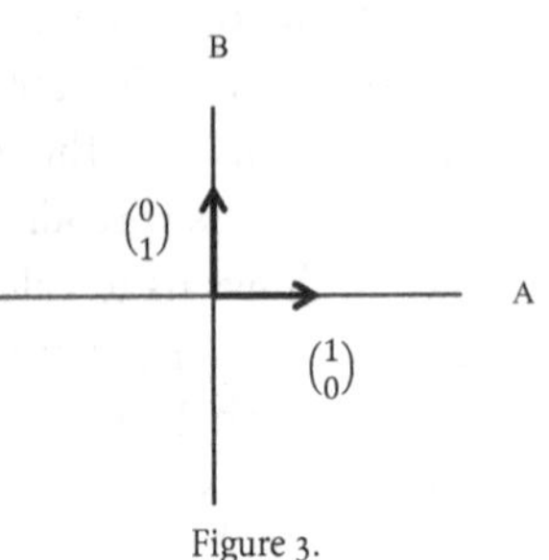

Figure 3.

The game of superposition, the idea that it is in some combined, *both A and B*, state seems like pure fiction. In the slit situation, you cannot give up on your *particle* concept, so you just say *it* goes through both slits, or *it* appears or acts as though *it* goes through both slits. However this simple two-state case and its simple 2-D representation suggests something more is possible. Suppose we concoct a new state in this mathematical world, writing it as figure 4. It can be shown on our A-B space as figure 5. Since A and B are perpendicular on our picture and are mutually exclusive in measurements, you get A *or* B if you do an A-B measurement, then why not consider two perpendicular possibilities $A' = \frac{1}{\sqrt{2}}\binom{1}{1}$ and $B' = \frac{1}{\sqrt{2}}\binom{1}{-1}$? Now we can make measurements in the A-B system or in the A'-B' system and superposition is not just a loss of knowledge, an inability to predict, an inextricable intrusion of probability. We find that these superpositions are possible and meaningful. That is, they are new and different *realities.* Making a measurement means a choice on the part of a measurer and the set-up of the measuring apparatus. Start with a pure state A (by finding a system which gives A when measured). Now measure in the A'-B' system. Our known state $A = \binom{1}{0}$ can be written as

$$\begin{pmatrix} \frac{1}{\sqrt{2}} \\ \frac{1}{\sqrt{2}} \end{pmatrix} \text{ or } \frac{1}{\sqrt{2}}\begin{pmatrix} 1 \\ 1 \end{pmatrix}$$

Figure 4.

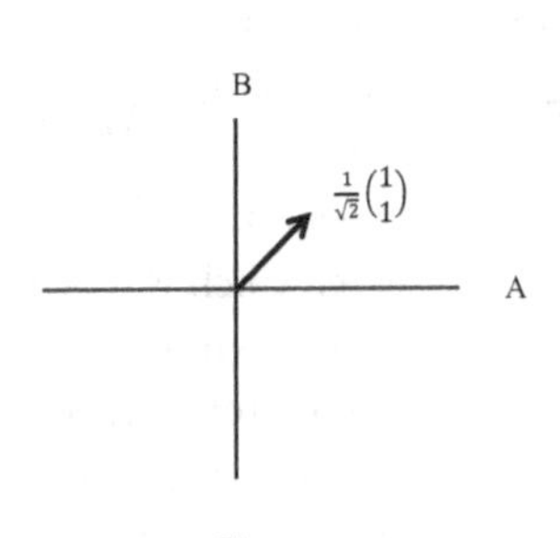

Figure 5.

$$\frac{1}{\sqrt{2}}\begin{pmatrix} \frac{1}{\sqrt{2}} \\ \frac{1}{\sqrt{2}} \end{pmatrix} + \frac{1}{\sqrt{2}}\begin{pmatrix} \frac{1}{\sqrt{2}} \\ -\frac{1}{\sqrt{2}} \end{pmatrix} \text{ or } \frac{1}{\sqrt{2}}A' + \frac{1}{\sqrt{2}}B' \text{ or } \begin{pmatrix} \frac{1}{\sqrt{2}} \\ \frac{1}{\sqrt{2}} \end{pmatrix}'$$

. Therefore we will get A' half the time and B' half the time. Making the measurement A'-B' on a pure state A forces it to decide: Am I A' or B'? and therefore does not allow it to stay in a pure state A. Whether it chooses A' or B', a subsequent measure of A-B will give only 50 percent A, not the 100 percent A which we started with. Thus a superposition state, e.g., a particle not exclusively here *or* there but here *and* there, is as real as an exclusive state. It *is* an exclusive state in a different system. Superposition states *extend* reality, requiring choice on the part of an observer and thus giving the observer a power of creation by the choice of measurement system. And states can be much more creative than our simple picture of a vector turning in any direction in our A-B plane; they may have phases which we include by complex numbers. I'll leave that for you theoretically minded explorers to sort out.

In case that is way too abstract for you, consider something specific we understand a bit. In the USA, we have a two-party system. Suppose it is all-encompassing and exclusive, i.e., *everyone* is either a Democrat or Republican, *if you ask*. Nobody is allowed *outside* that framework. But suppose we invent something different *within* the framework: Demlican and Repubocrat. These new categories can be also all-encompassing and exclusive. Every Demlican is half Democrat and half Republican and is clearly distinguishable from a Repubocrat which is also half Republican and half Democrat. We call this clear distinguishability "orthogonality" or "perpendicularity." The Republicans and Democrats meet on Monday night, in separate caucuses, of course, and the Demlicans and Repubocrats meet on Tuesdays. If you follow them from Monday to Tuesday, you will find half of each of the traditional parties show up with each of the new parties and there was no hint as to which will come out which way. And party loyalty never carries past the *most recent poll or caucus*. Physicists joke that we have particles on Monday, Wednesday, and Friday and waves on Tuesday, Thursday, and Saturday. One day a week we can rest from this nonsense; Jews, Seventh-Day Adventists, and other Sunday dissenters are free to make appropriate substitutions. I will leave it for you to define more creative parties for other days of the week, such as

REPUBOCRAT DEMLICAN

As we told you before, the Schrödinger *wave* mechanics and the Heisenberg *matrix* mechanics, coming from totally different pictures and approaches to science and looking totally different in the mathematics, are in fact mathematical twins having exact correspondences for every part. You will see books using the terms *wave* mechanics or *matrix* mechanics, maybe devoting chapters to each, but the more general term QM subsumes both. We usually use the wave form if the measurement of interest produces an infinite continuum of possible results; we use matrix mechanics if what we are measuring can take on only a few possible answers.

The key idea of all the quantum mechanical mathematical representations is to find a *state function* (variously called wave function or state vector) which tells us what we can know about the system. For any and all physical variables which we can measure, we need to find all possible values and the respective state vectors which will yield each value, thus representing pure states. When you know the state vector, you write it as some combination of pure states of the variable you want to measure. The probability of getting a particular value is proportional to the square of the coefficient of that pure state, and the probability is *all* you can get.

The special property of QM is that a superposition is a state which does not have a specific value but is a hybrid of the various states. With a coin, you get heads (H) or tails (T); with a quantum mechanical coin you may get a real state called 0.6 H + 0.8 T. When you ask if it is heads or tails, you will get 36 percent = $(0.6)^2$ heads and 64 percent = $(0.8)^2$ tails. But QM says it was not a *mixture* with 36 percent heads and 64 percent tails. Each of them in the state 0.6 H + 0.8 T was a different beast combining 0.6 head and 0.8 tail. Similarly, the electrons which approach a symmetric two-slit apparatus are *not* divided into the 50 percent which go through the right slit and 50 percent which go through the left slit. *All* of them go through both.

If one can calculate state functions as they change in time and know all the transformations for possible measurements, one can get the odds for all the specific values but *nothing more.*

QM may limit our *certain* knowledge of *specific* answers, even introducing randomness, albeit carefully controlled and shaped randomness. But it also expands the reality. A combination of A and B may mean that measurements of A and B are no longer certain, but in a new world at different angles in A-B space, there is a certainty in this new superposed world. The prototypical wave/particle duality opened a new set of realities. The God-man duality in Jesus makes it possible for us to participate in a divine-human duality. We are not stuck in the mortal flesh nor do we need to forsake it to achieve some relationship with the divine. Because the dualities of our faith and of QM are both mysterious, we can hardly suggest that either one illuminates the other, but they do have an interesting relationship. And what we believe and attempt to measure does make a difference.

KILROY WAS HERE

HEISENBERG MAY HAVE BEEN HERE

12

I'm Not Sure!

> Heisenberg's uncertainty principle gives clear quantitative limitations of our experience and knowledge, hopefully leading Christian thought away from the false clarity inspired by Newtonian mechanics into recovery of mystery and faith.

HEISENBERG HAD NEARLY FAILED his doctoral exam because of his lack of understanding of optical resolution. His embarrassment undoubtedly etched this basic theory on his mind and set him up for an understanding of QM which makes his name memorable, the eponymous *Heisenberg* uncertainty principle. This is doubly ironic in that the *wave* nature is the main feature, the explanatory key. When Heisenberg invented *matrix* mechanics and Schrödinger invented *wave* mechanics, the wave feature and the differential equation of waves, a calculus-based wave equation, were immediately more amenable to both visualization and calculation. Heisenberg, feeling the foundation of his new-found success being pulled from under his feet, was kicking and screaming against *waves*. The wave/particles are so unsettling that even today, over eighty years later, many physicists will think of an electron as a *particle* and light as a *wave*. Depending on one's point of view or the day of the week, one picture or the other will almost randomly appear to be the right way of looking at a problem. It is difficult to put the two together and that is what is necessary to visualize the uncertainty principle.

When particles go through a slit, they produce a pattern on a collecting screen which exactly images the slit (shown in figure 6); a particle goes through (where the slit is) or it doesn't (where the wall is). Thus we can collimate the beam of scattered particles (i.e., choose the ones along a narrow range of direction). There appears to be no limit to this process, in principle; narrowing the slit narrows the range of directions allowed through.

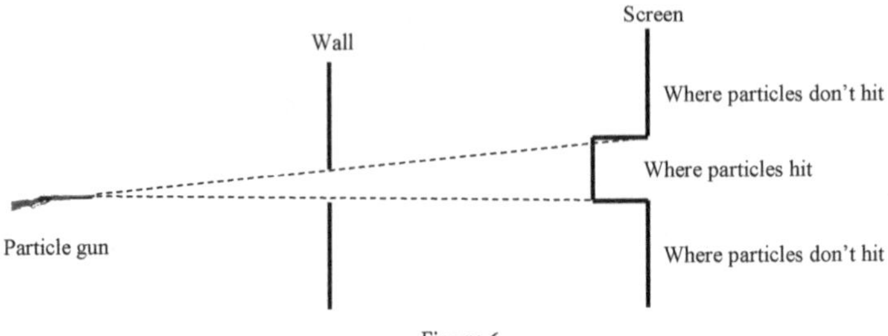

Figure 6.

With waves it is different. Well, eventually it is quite different even though there must be some ways that waves and particles behave almost the same. If the wavelength of waves is much smaller than the slit size, waves give the same result as particles. This is why a *camera obscura* (a simple dark box with a hole in it) can produce such marvelous results. The wavelengths of light are less than 1 micron (millionth of a meter) so that a 1 mm hole is 1000 wavelengths wide. In a large dark room a hole of several centimeters can produce nicely projected images. Trying to improve the picture (sharpen the edges) by making a smaller hole runs into two problems: 1) the image gets dimmer by the lesser amount of light and 2) eventually the wave-rays get bent (diffracted) when the hole size approaches the wavelength. The first is not a fundamental problem, just requiring more sensitive detectors or longer exposure times. The second is an unbeatable fundamental limit. At hole sizes of a wavelength (and smaller) the waves spread out everywhere and the picture is totally smeared out.

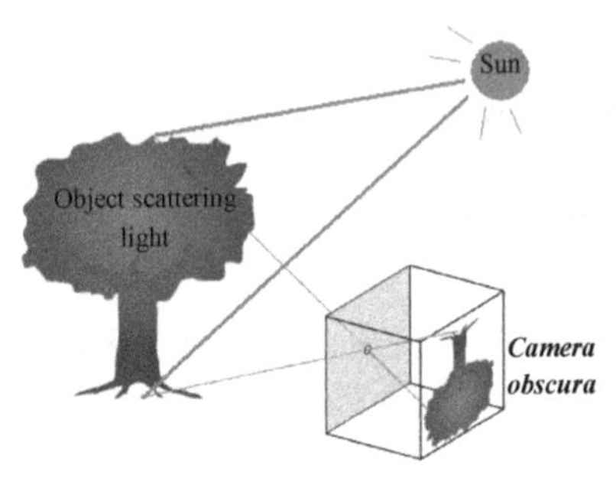

A large (much greater than a wavelength) object has a narrow (scattering) diffraction pattern which can *all* be caught in an optical system (with lenses) and the image will be faithful. As the object gets smaller, the image will get smaller proportionately as long as the optical system can collect the whole diffraction pattern. When the system can no longer collect it all, the image does not continue to get smaller. We call this *diffraction limited* and we

A picture taken with a *camera obscura*. Look on-line for many high-quality pictures.

cannot do better. Period. We, as humans, are limited to *visible* light, which has wavelengths thousands times the size of atoms and therefore diffraction patterns of light scattered from atoms go in all directions. Our eye can collect only a small fraction of the diffracted light and thus cannot see atoms appropriately sized; they look many times bigger and thus would be indistinguishable (to us) from very huge molecules. A single isolated atom can be *seen* (scattered light can be detected) but two atoms close together cannot be distinguished or *resolved*.

Heisenberg knew we could not visually follow an electron path inside an atom. Following the positivistic dictum of not talking about or using in your theory anything you cannot sense, he thus eschewed any idea of an orbiting electron. Nevertheless he achieved a theory which could not avoid asking questions about the position of the erstwhile particle called an electron. Although our human eyes can only detect *visible* light, a narrow band of the wide range of electromagnetic frequencies and wavelengths, Nature is not so limited. The Braggs had shown X-rays to be electromagnetic waves of wavelength about the size of atoms. Why not use shorter and shorter wavelengths to look with higher and higher resolution, finding the position down into the details of the atom? In principle that sounds fine; just make a gamma-ray microscope, gamma-ray photons having wavelengths one-millionth of that of *visible* light. We thus improve resolution (over the human eye) by a factor of one million, easily peering into the motion of an electron in an atom. But there was a catch, and that's why he used this illustration: the particle-like nature of electromagnetic waves says that the energy of a gamma-ray photon (the smallest piece of that radiation) is thus one million times that of a visible photon and thus kicks the electron to kingdom come and kingdom gone. If we try to find *where* it is to that accuracy, we don't know which direction or how fast it will be knocked. The wave/particle dual nature comes with a fundamental irremovable uncertainty.

The uncertainty was well-known in waves. The spread of a wave going through a hole (whose size is a measure of the range or uncertainty of the position) was in inverse relation to the size of the hole; smaller hole means

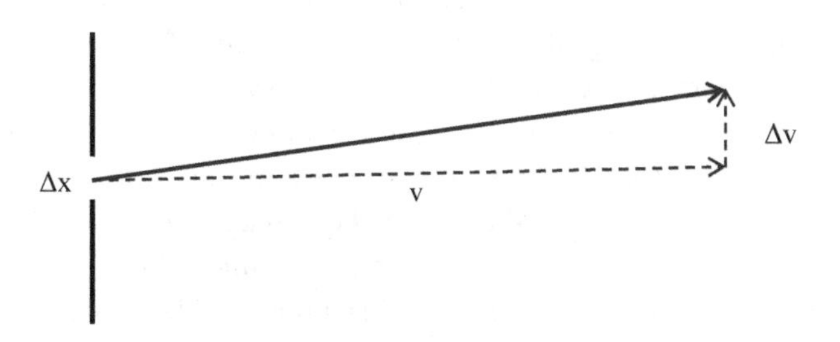

larger spread. Δx is the uncertainty in the (up-down) position. Δv is the uncertainty in the (up-down) velocity v. The basic relationship is

$\frac{\Delta v}{v} \sim \frac{\lambda}{\Delta x}$, ~ means "roughly equal," λ means wavelength.

Translating to particles, the uncertainty in (up-down) momentum (compared to the total momentum) is

$$\frac{\Delta(\text{Momentum})}{\text{Momentum}} \sim \frac{\lambda}{\Delta x}$$

$$\Delta(\text{Momentum})\ \Delta(\text{Position}) \sim h$$

because (Momentum) (λ) = h by de Broglie's conversion between particles and wave properties. The well-known wave phenomenon of diffraction (spreading or bending) from a single slit becomes an uncertainty relationship for a single particle. We have said this in rather crude simple relation-

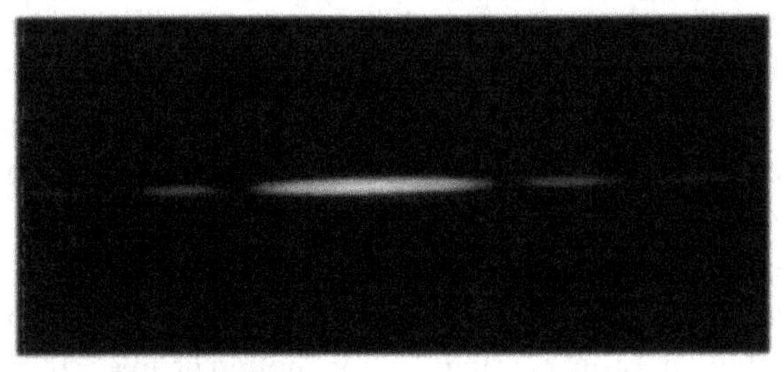

Diffraction (bending, spreading) of light from a narrow vertical slit

ships, but the theory of waves and Fourier expansion (which Heisenberg was using to relate what he could see and measure, frequencies and intensities of the light emitted from atoms, to what he could not see, electron motion in the atom) allows complete definitions and quantification of these crudely defined terms. What we find is that uncertainties can be large without limit (nobody limits your ignorance) but there is a limit on your knowledge or your ability to prepare a particle. There is a strict *inequality*: $\Delta(\text{Momentum})\ \Delta(\text{Position}) \geq h/4\pi$. Δ here has a well-defined statistical meaning, standard deviation.

The significance of this is absolutely earth-shattering to the world of classical mechanics. The supreme statement of classical physics is the claim that a Supreme Intelligent Being can know the complete history and future of every moving piece of the universe by solving Newton's equations of motion *if* the position and momentum are known exactly at one time. We must start from a *point* in x-p space (i.e., position-momentum space (Fig. 1 below), but QM is telling us that the smallest area in x-p space is $h/4\pi$ (Fig. 2). A *point* cannot be known, or prepared experimentally, or meaningfully exist. The process of finding or preparing a point is like a toothpaste tube:

squeeze in one direction and the fluid forces the tube out in the other direction (Figs. 3, 4).

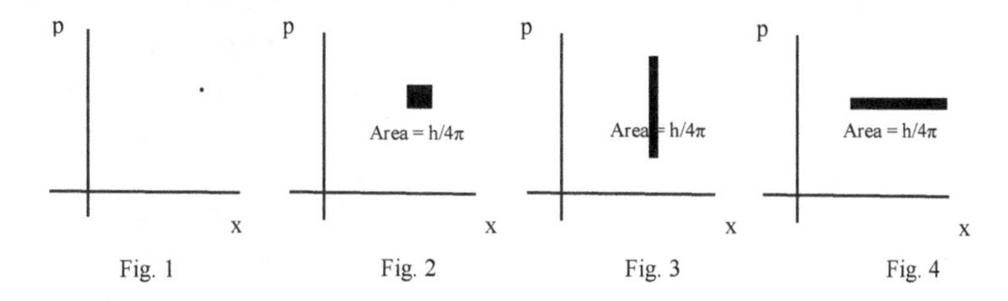

As we know, the units of h are not only Momentum times Position [(ML/T)(L)] but also Energy times Time [$(ML^2/T^2)(T)$]. Thus we should expect and do find a similar uncertainty principle with energy and time. With the famous Planck-Einstein-de Broglie relationships of Momentum = h/λ and Energy = hf, we can see that the factor h can be factored out and two dimensionless uncertainty principles can be written

$\Delta(h/\lambda)(\Delta x) \geq h$, which implies $\Delta(1/\lambda)(\Delta x) \geq 1$.

$\Delta(hf)\ (\Delta t) \geq h$, which implies $(\Delta f)(\Delta t) \geq 1$.

The inequalities on the right side are classical *wave* relationships, independently of QM and preceding QM; the left side inequalities are the unavoidable QM equivalents for particles. The Δt (the limited range in time) can be accomplished like a slit in space; that is, we can make a "slit in time" by allowing a wave only for a brief time. Start with a pure frequency, looking like this in time (with no beginning or end, dashes indicating going to infinity). Turn it on for a brief time Δt. We say that it is not a pure frequency

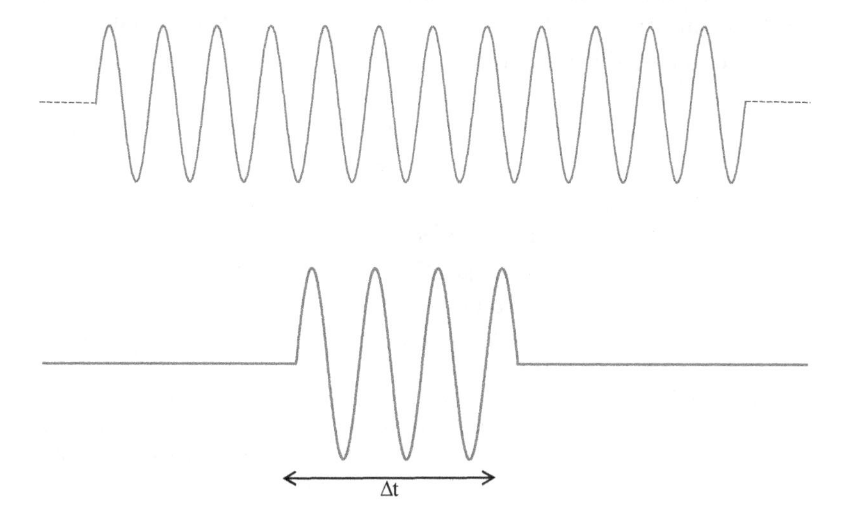

anymore. Sound is a good example since we hear frequency as pitch. Mathematically we say that this shortened tone is made up of a range of frequencies, something like this. When we hear such a brief tone, we lose the ability to distinguish pitches. In fact, it will no longer sound like a tone (a pitched note) but will sound like a click (an unpitched noise). Nowadays this can be done with digital sounds;[1] not many years ago, I did it by cutting small pieces of audio tape of a steady tone and attaching them to a blank tape. The relationship $(\Delta f)(\Delta t) = 1$ is a limit. It is possible to play a whole range of frequencies (Δf much greater than 1) for a long time (Δt much greater than 1) (we call that white noise), but you cannot make Δf small (well-defined pitch) in a short time. Very low bass notes produced for short times (e.g., pizzicato or a slap on a double-bass) do not have to be, cannot in fact be played in tune exactly.

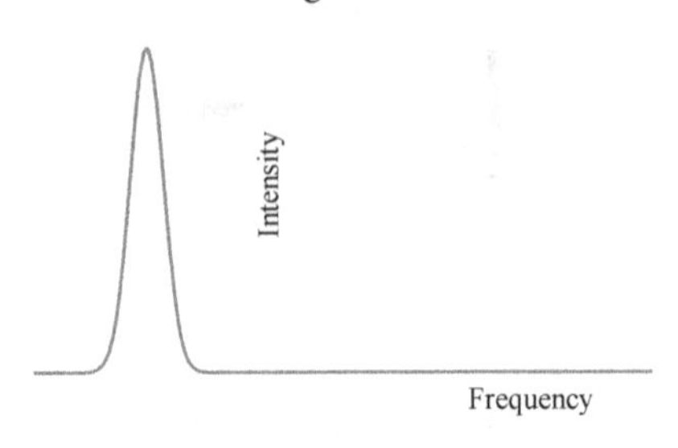

Heisenberg, perhaps eager for revolutionary conceptual change, saw this breach in the classical physics fortress as the end of a favored foundational framework of physics, causality. Time, space, and causality were intricately woven into our conception of the world and this looked like an ultimate limit. With real world sizes and speeds, like baseballs as fast as we can throw them, the infinite accuracy theoretically seen in calculus does not seem very threatened when the uncertainty principle questions the twentieth decimal place and our best practical measurements give us only six to ten decimal places or significant figures as we call them. But in the atom, an electron now sees uncertainties as big as the quantities. The toothpaste tube bursts and all semblance of a law of motion is gone. Nobody should speak of an electron orbiting the nucleus in an atom, but we don't know what else to say.

QM extends the uncertainty principle to all kinds of situations. Every measurable variable is represented by a mathematical operator. When two operators do not commute (i.e., the result depends on which one comes first), we call them incompatible and an uncertainty is born. The product of uncertainties depends on how badly they don't commute, i.e., on how big is the difference between PQ and QP. And you can guess, I hope, that h, that very small but unavoidably *finite* Planck constant, always shows up. You can

1. Physics departments often have auditory demonstrations on-line for hearing such effects. A CD of *Auditory Demonstrations* was produced by A. J. M. Houtsma, T. D. Rossing, and W. M. Wagenaars and is commercially available. The effect described here is illustrated on Track 13: "Pitch salience and time duration."

also reasonably guess that both calculus (the basis of Schrödinger's wave mechanics) and matrix algebra (the basis of Heisenberg's matrix mechanics) have mathematical operators which do not commute.

In the language of the last chapter, two incompatible operators cannot have the same pure states. A pure state of P means that the result of measuring P is fixed and knowable. Pure states of either P or Q can be prepared with the specific values attached to that pure state. When P and Q are incompatible, a pure state of P is necessarily a superposition of pure states of Q. A superposition state allows us to know the possible answers and the odds for each one, but no one has a hint as to how to break the code and determine which value will come up. The theory is so good at what it tells us and so opaque for seeing further that we believe that the individual results carry an inscrutable randomness. Note the word *believe*; it's also *my belief* that this question will remain unanswerable.

Where does this uncertainty principle leave us? If it and our interpretation of it are correct, it clearly charts very real limits to our knowledge. Because what *is* and what is *knowable* seem to have a close connection, this seems to place very real limits on how the world *is*. We find first-rate physicists daring to use the word *mystery*. We have given up on the simple metaphysics of classical physics, a world where what is seen is what is and what is knowable. There is something beyond physics and it's all right to have limits and it's also all right to do our best to see behind the veil. The American Protestantism that I have seen has tried so hard to follow Newton's presumed success of perfect laws with no mystery. We are so damned concerned about being right that we cannot use the word *mystery* except in some emaciated form: Was it Colonel Mustard in the den with a lead pipe? It is no wonder that science has been given such high place in the philosophical, religious, and spiritual aspects of our lives. It has perhaps faced the mystery more squarely, sought the mystery more honestly, parsed the mystery more clearly, and believed the mystery more deeply than our churches, our seminaries, and our philosophy departments. I am not beatifying science. I spent more time in science classes than in Sunday School, and I know that the party line in science is as rigid and over-simplified as an average Christian's belief system and as childish as the litany of Bible stories. But both science and religion are marvelous back doors of the wardrobe which opens into another world, a world of mystery and faith, a world no less real because of its uncertainties, even your uncertainty of its reality.

I'm not sure! But I believe!

13

Complementarity: Either . . . Or . . . , Nothing But . . . , Both . . . And . . .

Bohr, the leading interpreter of QM, saw complementarity, a necessary reality of two apparently contradictory concepts, as foundational for all knowledge; it is perhaps better used in theology and the practice of our faith.

"Even as empiricism is winning the mind, transcendentalism continues to win the heart. Every major religion today is a winner in the Darwinian struggle waged among cultures. . . . People need a sacred narrative. They must have a sense of larger purpose. . . . The essence of humanity's spiritual dilemma is that we . . . accept[ed] one truth and discovered another. Is there a way to erase the dilemma, to resolve the contradictions between the transcendentalist and the empiricist world views?"

E. O. Wilson,[1] sociobiologist, not known as a friend of religion

"We . . . comfort ourselves with . . . the old saying of the two kinds of truth: . . . one . . . so simple and clear that the opposite assertion could not be defended; the other . . . so-called 'deep truths' are statements in which the opposite also contains deep truth."

Niels Bohr[2]

1. Wilson, *Consilience,* 286, 267, 289.
2. Bohr, *Atomic Physics and Human Knowledge*, 66.

> "Complementarity represents . . . the most revolutionary conception of our day."
>
> John Wheeler,[3] quoted by Alfred Landé

> "Created matter and uncreated divinity complement each other and have to be combined because of this drive toward unification in God's triune nature."
>
> Christopher Mooney[4]

FAITH IS THE EVIDENCE of things unseen. Faith, that one-word descriptor of religion, is thus a paradox, a contradictory statement. One does not have to be a linguist to see that the word *evidence* has that root "vid" which we know from *video*, "*veni, vidi, vici*," etc. *Evidence* means "something obvious to the eye or the mind." Yes, people of faith are simply saying that they see the unseen and you can laugh if you wish. Faith is not believing against your better judgment or fervently screwing up your mind to imagine something is true when it is clearly false. Faith is to have eyes that work in regions beyond the visible, and not just the trivial sense of ultraviolet, infrared, and the wide range of frequencies which our physical eyes are not capable of detecting. Faith steps past the boundaries of the physical world, knowing there is something there, tapping into resources that natural sciences have declared off-limits or non-existent, a region where belief is part of the price of entry. Belief is also a necessary ticket for the physical sciences, but present-day scientists imbibed that belief subconsciously as part of their cultural heritage; they often also imbibed the spirit of the age which can smugly look down on "those who were . . . exposed, in youth, to this particular virus [religion]."[5] I will admit to gratefulness for being exposed to both viruses, religion and science, in my youth; I will also declare that the cheap Sunday School version of religion taught to children has its counterpart in the cheap and weak *science* program for children, certainly through college. They both have their weaknesses and can easily serve as a vaccination, immunizing us against further development. I'm afraid most of our culture becomes immune to both; many people of science are immune to religious faith and many people of religious faith lose any scientific faith or distort it into a supposed handmaiden for their physical *and* spiritual needs. A supposed scientist of religious faith, which many of us try to be, is seen as an oxymoron, an incarnation of paradox.

3. Landé *From Dualism to Unity*, 81. Almost identical quote in Petruccioli and McGilvray, *Atoms, Metaphors, and Paradoxes*, 30.

4. Mooney, *Theology and Scientific Knowledge*, 168.

5. Sokal, *Beyond the Hoax*, 427.

In the beginnings of Western philosophy, mathematics had the place of the other-worldly *real* while matter was less real, serving as a prison for our eternal souls. Pythagoras had almost made the integers into gods, each number having symbolic, mystical meaning. The basis of music was the relationship between notes produced on a *whole* string, *half*-string, *third*-string, etc., frequencies in ratios of 1:2:3:4:5 . . . . A ratio of 2:1 is an octave; 3:2, a fifth; 4:3, a fourth; 5:4, a major third; 6:5, a minor third; 9:8 and 10:9, whole steps; and 16:15, a half-step. (You may note some missing, for which you may need to go to jazz, blues, or some other part of the world.) These set a standard for *harmony* and *consonance;* deviations produce *disharmony* and *dissonance.* The mathematical designation of "rational" for numbers which are ratios of integers and "irrational" otherwise carries that Pythagorean near-worship of integers into our modern language where no one wishes to be considered *irrational.* There is a story that the discovery of the irrationality of the square root of 2, the indisputably naturally occurring length of the hypotenuse of a right triangle of sides 1, resulted in murder, mass suicide, or at least a mysterious disappearance. With Plato, an admirer of Pythagoras, geometry was added as the presumably perfect mathematical touchstone and admission requirement to his, the first university of Western civilization. This blossomed into Euclid's geometry, the epitome of deductive certainty and the proto-textbook of mathematics. Calculus, with its use of the infinitesimal, had dared to veer into paradoxical uncertain waters, but attached its unlimited precision to analytic geometry and the commonsense ontology of classical mechanics to flip the Platonic idealism on its head.[6] The material, the mundane, was the certain, the real, the lawful and gradually held the upper hand, having no place for the celestial, the ethereal, the eternal, the ideal. Whereas mathematics could almost be considered the religion of the Greek Golden Age, an unquestionable source of the West's intellectual tradition, modern science, seventeenth-century physics, co-opted the mathematics and became the standard, the ideal of knowledge, the new exclusive religion. Christianity could not avoid adoption of the Platonic and thus took sides in the conflict between ideal ideas and the messed-up mass of material, between celestial and mundane, between rational and irrational, between perfect and immortal being and the woeful becoming of this world. In seventeenth century modern science, the faux heavens of the mathematical world joined forces with the faux lawfulness of material motion to at least project the image of complete knowledge, *scientia.*

With the twentieth century, physics began to repay a debt to the Judeo-Christian faith's serious and positive view of the physical world. A faith

6. Alexander, *Infinitesimal.*

which says God made the world, declares the world to be very good, sustains the world, enters the world, and has great plans for the world does not disdain the seen. Physics, from its seventeenth-century rise, had separated the physical world from God and thus made great progress in understanding and explaining the great machine of the physical world, gradually stifling the unseen world and the unseen God, until God could by the commonsense rules of the reigning science, be reasonably declared dead, no malice, no murder, just no longer needed as an hypothesis. But science also sowed the seeds of its own demise, at least as the simple reigning sovereign over the entire world.[7] Its imperial designs kept pushing the limits of time, space (outer and inner), speed, substance, and concepts. Two limiting numbers were discovered, each wrecking the commonsense realities. The velocity of light, c, appeared as a speed limit of the universe, limiting our commonsense views of space and time to speeds much less than c, and simultaneously affirming an absoluteness as a basis for Nature and natural laws. Planck's discovery of an inner limit, a limit on the small side, was quantified by the constant of action h. This very small number, roughly 10^{-34} when expressed in units of laboratory sizes (kilograms, meters, and seconds), assured us that there were limits to our commonsense ideas and our "seeing," the world of pictorial representation, causality, and determinism; simple pieces of matter would work as we knew them only in the realms with action much bigger than h. Action, probably not mentioned in any physics course you might have had, is measured as the product of momentum and distance or of energy and time. The action of basic small particles, the electron, at the atomic sizes, gets exactly into the range of h. We would in physics say it in the opposite direction: the size of the electron's mass, the strength of electric pull on the electronic charge giving an electron energy and momentum, and our limiting constant of action *determines* the size of atoms, saying that you cannot squeeze that "mostly empty space" any more. Thus the faith of physics did not stop at the limit of the "seen." It successfully ventured past the boundary into the "unseen." Consequently it partook of all the paradox of religious faith and the difficulties of such a language. A person outside both physics and Christian faith can apply the same skeptical doubt to each; I have experienced both brands of skepticism in a single person, expressed in either order, depending on which of my identities came to light first.

We may say that in QM all roads lead into or out of Copenhagen, the kingdom of Niels Bohr. A small man, befitting the serious soccer player he was, from a small country, he was nevertheless the towering figure of quantum physics, the spokesman and leader. Einstein was the believer in law

7. Rosenblum and Kuttner, *Quantum Enigma,* 87.

and order of an impersonal God, Nature, who brooked no contradictions or paradoxes; Bohr, with no more claims of a personal God, nevertheless had "been arguing with Einstein all the time,"[8] whether Einstein was present or not, dead or alive. In a manner typical of Bohr, he had (not unquestioningly) accepted what he knew had to be: the world is full of paradox and contradiction and the paradoxical picture had forced itself into physics, only so recently the hopefully antiseptic paradox-free classical world.

Kierkegaard, the early nineteenth-century existential philosopher, had declared that paradox was at the heart of Christianity. I wish I could give you a detailed analysis of Kierkegaard's influence on Bohr's thinking; I doubt if Bohr himself could give such an account.[9] Such effects can be very subtle, our philosophical moorings being often deep but hidden, especially when covered by the hard coat of scientific training. The possible influences are known. Bohr speaks, in a letter to his brother, of a specific book by Kierkegaard, but never avows any specific influence.[10] More importantly, the long-time family and personal friend, Harald Høffding, was a philosophy professor at the University of Copenhagen, including Niels Bohr among his students, and an expert on Søren Kierkegaard, writing a book on Kierkegaard as philosopher during the childhood of Niels Bohr. The children, Niels and Harald, were allowed to sit in the discussions of the small group of professors of the University of Copenhagen which included Høffding, the elder Bohr, and a physics professor of Christian faith. Niels's father was a professor of physiology and struggled through the controversy between mechanism and vitalism with his friend Høffding. This certainly sensitized the young Niels to the issue of seemingly opposite approaches in the wider realm of knowledge, an issue which he returned to explicitly in his later writings albeit with no great depth. For Einstein, paradox was only a goad to strive for a new foothold, providing at most a provisional scaffolding, no place to rest for long. For Bohr, paradox became complementarity, a dreamed-of foundation for a grand palace, the unity of knowledge. It was a building which Bohr never came close to building, but he thought the

8. Bohr, *Atomic Physics and Human Knowledge*, 66.

9. Later in life, Bohr certainly felt a kinship with William James's work, but there is disagreement about when he became aware of James. Stapp claims that Bohr did not know James's work until the 1930s, working out some similar ideas independently. Stapp, "Copenhagen interpretation," 1115.

10. Margrethe, Niels's wife, while affirming Niels's love of Kierkegaard's writing, even his gift of a specific book to her, disavowed his interest in the questions and thinking of the great Danish philosopher. *AIP*, Oral interviews. We catch glimpses of an alternate view, e.g., from Bohm's correspondence: "[Bohr] admits (privately) that he was strongly affected by Kierkegaard's Existentialism." Bohm and Biederman, *Bohm-Biederman Correspondence*, 8.

foundation was there in QM, in spite of Einstein's pesky refusal to accept it. A QM built on complementarity was not the real McCoy (or the *true Jacob*, as Einstein's Swabian phrase is literally translated).[11]

Complementarity may have been one of the tap roots of quantum mechanical interpretation, but it is one which has gone almost completely underground, the word hardly appearing in modern textbooks in the field. Bohr's notoriously bad public communication, so easily forgiven by those who knew him personally, leaves this supposed mainstay as a vague and poorly developed concept. His writings and speaking on the topic are primarily in ceremonial addresses by the ageing hero of physics at professional meetings of non-physicists. They are typically apologetic about Bohr's own lack of depth in the field of his listeners and reassuring them about the "epistemological lessons" of physics, as though the forced broader view of complementarity with renunciations of old classical concepts had been given a rigorous mathematical and interpretational basis which could prove useful in other disciplines. "By the ingenious formalism of QM . . . it has actually been possible along the lines of complementary description to bring logical order within a rich field of (physical and chemical) experience."[12] He puts great faith in the "elegant axiomatic exposition [of QM] of von Neumann."[13] Einstein was never impressed and it is not clear what other disciplines should gain from physics' experience other than the comfort that physics was not as simple and clean-cut as the old clockwork mechanism indicated. But that is no small thing. The narrow approach of the old physics was inadequate and, at least by analogy and metaphor but not by axiomatic exposition or abstract mathematical formalism, other disciplines could profitably quit following the old physics paradigm, or quit being tempted to follow the old physics.

Wave/particle duality represents what might be called a narrow conception of complementarity. Because of the battles about the reality of each of these aspects in both light and matter and their key role in the development of QM, this is a form of complementarity, usually called duality, which is still prominent in the teaching of QM. The particle view of light, photons, could not adequately represent the interference phenomena so well explained by electromagnetic waves. A synthesis of two mutually exclusive entities/conceptions, each of which could be applied separately under various conditions, was necessary to obtain a comprehensive account. This

11. Quoted in Kaku, *Einstein's Cosmos*, 127. It is interesting that the name *Jacob* means "deceiver" in Hebrew.

12. Bohr et al., *Collected Works, Vol.10*, 70.

13. Bohr, *Philosophical Writings*, 83.

synthesis involved, at least for Bohr, a necessary renunciation of determinism, causality, and the mindless carry-over of the commonsense reality of classical physics into the atomic world. Light acts as a wave or as a particle depending on the kind of measurement one performs. A weaker, but perhaps richer, concept of causality and reality are necessary because of the dual nature and the dependence on the observer's choice. Biology, with the conflicting approaches of mechanism and vitalism, was close to Bohr's heart because of his father's involvement in physiology. To Bohr, they were complementary/dual approaches. Mechanism represented only one approach. Vitalism, in spite of his great concern to avoid concepts of a mystic life force which might go counter to the spirit of science, was also an indispensable approach. The holistic and finalistic aspects of biology could certainly not be handled by reduction to atomic physics. Just as the quantum of action had limited us in physics and forced a dual viewpoint, the very existence of life was a fundamental fact, forcing a complementarity in biological explanations. Life, as Bohr knew well, is often falsely conceived as counter to the second law of thermodynamics; life creates order while thermodynamics predicts the heat death of the universe, both processes giving a direction to time which does not exist in classical mechanics and electromagnetic theory. The contradiction is a false one because the second law of thermodynamics deals with a closed system (which may be supposed for the universe as a whole) while life cannot be, is never, a closed system. The necessary exchange of matter and energy by a living organism with its surroundings means that a living organism cannot be treated as a mechanical fixed set of atoms; life requires more than physics. Biology thus requires the holistic, purposeful organismic view that has been mercilessly wrung out of the Aristotelian view of mechanics, leaving the desiccated Newtonian mechanics as a supposed model for all knowledge.

Part of the cost of QM had been the loss of causality. But our old form of causality had been raised to its high position by the loss of purpose. Aristotelian science had four causes: *material* causes (the substance), *formal* causes (shape or form), *efficient* causes (what we might call immediate or secondary causes such as forces in mechanics) and *final* causes (purpose or *telos*). Classical mechanics had narrowed *material* cause to a single attribute, the mass or inertia of matter; the *form* was unimportant as masses came to be treated as though they were points (at the center of mass); *efficient* causes were reduced to a single concept, the force; *final* cause or purpose was lost, the unneeded hypothesis.

Classical physics as developed from the seventeenth to the nineteenth century had elevated its narrow causality, a strict unyielding causality for motion, to the position as an ideal rational explanation for all knowledge and thus caused a chasm between science and religion. Not only did science develop the unbreakable laws of Nature as a substitute for manifestations of providence, it also narrowed its methods and viewpoints to preclude consideration of the human values and ideals which were necessary in a religious framework. They went fishing with a net that would not catch the God-fish. From this framework, we developed a distinction between objective knowledge and subjective belief. The loss of the narrow causality that had wreaked such havoc should probably not be considered disastrous, especially if we regained some complementary pairs such as cause and purpose or science and religion. Although Bohr decries the schisms and the losses inherent in our classical physics, he stops short of any clear call to return to purpose and/or religion.

In what Bohr considers to be a broadening and deepening of complementarity and its possible use to other disciplines, he emphasized the irreducible role of the measuring instruments, which ultimately involves the observer, a conscious decision maker. Einstein had vanquished the old concept of an absolute space and time by noting that observers who were moving with respect to each other measured different times and lengths. Length and time were not absolute; they were relative. We could no longer speak of an absolute space or time, but only of a time or space as seen by a specific observer. Bohr now saw the quantum revolution as going one step further. One could no longer speak of any description or conception of the *real* quantum object as it is independent of the rest of the world; the phenomena available to us for quantum theory are *only* the interactions between the quantum world and our measuring instruments which live in our world described by classical physics. He is led to a radical epistemology: we can know only the reality of our macroscopic classical world which has very clear real effects due to the unseen, unknown quantum world. All we can know of that inner world is relative to the measurement set up by experimenters.

Einstein's relativity gave up absolute space and time but clung tightly to absoluteness in two forms: the absolute speed of light in vacuum and the absolute form of laws of physics. Neither of these was relative. It appeared to Einstein that QM had failed to retain the essentials; it had given away the store. Causality had given way to an intrinsic irreducible probabilism, one with strict statistical laws to be sure. Reality of the fundamental pieces had been reduced to their effects in *our* reality. And those effects were so shaky as to produce a fuzzy image of a two-sided reality, two sides which could not be seen at the same time.

The effects of the measuring instruments were quite apparent in the broader world, certainly not needing physics to teach it. Any science which tries to look at people (past their trivial physical manifestations) runs into essential paradoxes of the object/subject relation. Bohr liked to remind us that the great religions always knew that we are both spectators and actors in the great drama of life. Now that physics had arrived at this glimpse of a new world (new for physics), Bohr saw possibilities of a unity of knowledge, knowledge which had been torn asunder by an admittedly successful mechanistic physics which had broken away from religion and any taint of purpose and metaphysics and was pulling the rest of knowledge with it. A suspension of quantum-level causality, representation, and visualizability was too much for the old school; Schrödinger called it "a doctrine born of distress."[14] Bohr did not see the intricate mathematical formalism as describing a causal evolution of real properties of quantum systems, but only the evolution of probabilities. It was a *symbolic* mechanics. Bohr reached much further back for symbolism and wisdom when he chose symbols for his coat of arms upon his award of the Order of the Elephant by the Danish government: the famous complementary wisdom of China in the yin-yang symbol (*taijitu*) and the Latin motto *contraria sunt complementa* (opposites are complementary).

Without using the word *complementary*, Donald McKay[15] nevertheless used the concept to confidently defend the Christian conception of soul and spirit against the scientific mechanism. It has been popular to declare a complete scientific reduction of the human being to chemistry and physics. McKay says: let them try; they can only achieve a trivial reduction; it is a fallacy of *nothing-but*tery. Mathematicians may fill blackboards with chalk, something which may be analyzed as *nothing but* calcium carbonate layers on a dark surface. But the meaning is in equation and words, at a level quite distinct from the physical. If our minds are thoroughly analyzed by the physics and chemistry, it still requires another level to speak of our function and meaning as persons. We do not need physics to know that complementarity is necessary. Physics is a Johnny-come-lately, but for some it gives the principle added authority.

The culture had accorded physics so much authority that philosophers tried to adopt the big names in physics as support for their respective school

14. Plotnitsky, *Principles of Quantum Theory*, 86.

15. McKay, *Christianity in a Mechanistic Universe*, 56–58.

of philosophy; Bohr's name was thus invoked in support of essentially all brands of philosophy. In spite of his closeness to Harald Høffding, he generally distanced himself from philosophy and philosophers. Although complementarity wormed its way into the framework of QM, almost no one, including philosophers, took up the principle as a basis for a grander vision. It would seem that theology, especially a theology that could see paradox as central to the Christian faith, would have been a better prospective ally for Bohr. His superficial involvement with Lutheranism (a baptism of social convention in his childhood somewhat akin to my baptism into six-man football upon moving to small-town Texas in junior high) and with Eastern religions (adoption of the yin-yang symbol for his coat-of-arms) did not promote what could have been a solid venture of some depth into a complementarity of religion and science or the complementarity within the much more difficult and treacherous ground within communities of faith.

Not being savvy in theological matters or even in the less substantive matters of Christian practice, I cannot do justice to the possibilities of complementarity, duality, and paradox in theology and the practice of faith. Just as wave/particle duality naturally crops up in the practice of QM (even though the sophisticates may tell us that the duality is unnecessary and a sign of our lack of development in quantum field theory or some other approach to the foundations, e.g., Polkinghorne[16] and Landé[17]), analogous dualities are suggested throughout areas of common Christian practice. These should probably be considered superficial complementarity. The deeply paradoxical aspects of God probably deserve the half-joking Bohrian aphorism of "deep truth" (quoted at the chapter beginning). Kierkegaard boldly declared that paradox was at the heart of Christianity. His clear attack on the institutional state church of Denmark has left him as a prophet without honor in his own country in spite of his honor outside Denmark and the recognition of his elegant language and story-telling in a country known for Hans Christian Anderson and Karen Blixen. The unsettling tendency of the church to divide may reasonably be considered a result of both shallow and deep truths which have two sides, sometimes apparently contradictory and exclusive, both sides worthy of adherence. Most of us are incapable of holding such apparent opposites in tension without taking sides. In the brief introduction to "Christian complementarity" that follows, my own prejudices or leanings may sometimes show through; I expect you too will have certain proclivities.

16. Polkinghorne, *Quantum World*, 7. "Since that day of Dirac's discovery the dual nature of light as wave and particle has been free of paradox for those in the know."

17. Landé, *From Dualism to Unity*.

The specific particularities of God's interaction in the world and the express biblical intent of a perfectly general interaction with the *whole* world is like a hurricane of duality which spawns multiple tornados of the same character and even a matching intensity in a local situation. The specificity of Israel as God's chosen people and of the Incarnation in one man/God Christ Jesus is a scandalous particle-like one-way street even though both manifestations are also explicitly to bless *all* nations in the covenant and blessing of Abraham and to save the *whole* world in Christ Jesus. Jesus comes *first* to the people of Israel but the Holy Spirit speaks *all* languages on the day of Pentecost. The absolutism of science is a bottom-up reduction which spreads its tentacles to everything; the absolutism of faith is a top-down inclusion which similarly covers all. The polar opposites act like impulses from both North and South poles reaching to all points on our spherical Earth. Our scientific and religious poles, the inaccessible world of the infinitesimal and the inaccessible infinite God, are like a set of fixed points spreading their influence in sometimes contradictory terms to this merry-go-round of life.

It appears to me that Kierkegaard is right that paradox is at the very heart of Christianity. This is a natural outworking of the transcendence of God. We believe firmly that God, in his grace, has made himself accessible to us. He has come to us; he has made himself known. *But* he is also inaccessible, beyond our understanding. Just as the common scientific mode tells us there is one truth and there is one method of getting there, there is a religious mode, usually in a fundamentalist version or among preachers who consider themselves the authoritative interpreters of the revelatory scriptures, making exclusive truth claims. There is a big difference between staking claims on the *existence* of an absolute Truth and on the *possession* of that absolute Truth. It may be that deep down, in hope and faith, all of us are committed to one ultimate meaning, reality, and truth and we commit our lives to our best discovery of that truth. Science has now given us at least a metaphor of that inaccessibility, of the paradoxical nature of our best efforts, of a limitation in our knowledge. We should not have needed science to confirm this; church councils struggled over such polar dualities in the nature of God and in Christ many hundreds of years before physics broke the stronghold of a narrow classical conception of truth which so much of the church was all too ready to imitate. The mystery of God was wrung out of the church to the extent they believed science methods and truth claims had wrung the mystery out of Nature. We must know that we don't understand and we can fruitfully try the models of science's limitations in its knowledge of Nature as possible models for our knowledge of God. They may at least be shadows of the real thing.

If apparently contradictory aspects are needed to give a complete account of our experience, we have needless debates and divisions. Creation/evolution is one of those fruitless and unnecessary debates. Creation is a pillar of our faith, arguably the first step in any theology which includes the world. But that does not mean its counterpart, evolution, the other pole, is wrong. Evolution is creation where we think we understand a little of it. We might even say that evolution is creation revealed. In that sense we might say that, in the Godhead, Jesus the Son is the analogue for evolution while God, the Father, is the analogue for creation. Genesis 1 presents the majestic, imperial (and maybe inaccessible) creator, while Genesis 2 presents the God who walks with his created ones. We don't *see* God and we don't *see* creation in the narrow sense of the beginning; we do see evolution and the Son as our earnest money. Science then appears as an analogue of the Holy Spirit. The Spirit reveals Christ, the Incarnate Creator, as science reveals evolution, the incarnation of creation. Jesus was God Incognito; evolution is creation incognito. For those of you who learned to see evolution as the enemy, I can understand this is way too much to accept. That's all right.

The creation/evolution controversy may rightly deserve the attention of numerous tomes, but it also deserves to be seen as a simple analogue of the particle/wave duality. It represents a complementary pair, perhaps mutually exclusive in conception, which are both necessary for a complete description. The narrow conception of creation limited to the beginning point in space and time is clearly particle-like. Evolution is wave-like, spread continuously throughout space and time, a continuous sustained creation. Creation limited to a beginning is *de*istic, not reaching the level of *theism* of a sustaining and active God.

Almost every idea that I entertain about my spiritual life or how God relates to the world can be seen in the light of this form of duality. I would generally argue against a division into Nature and Supernature. Nature is Supernature Incognito. But the manifestations of Supernature, which we often call miracles, are generally events, particle-like, while Nature is the wave, whirling continuously all around us. The propagation of the QM wave function, Wheeler's great smoky dragon (see next chapter) filling every nook and cranny, clouded by the smoke, represents God's invisible presence which pervades and upholds all of Nature. Then unknown in his timing or place, he bites and we see the particle manifestation, a miracle. These miraculous events, for example, a miracle of healing, given meaning by timing or extraordinary quality, are really no more or less miraculous than the garden-variety everyday good health that God blesses most of us with most of the time. New birth, being born again or being born physically for the first time, is a special event, but one that in no way can exclude or eclipse the amazing work of grace and health which sustains a long-term development of that life.

There is a branch of the Christian church (usually associated with the adjectives "revivalist," "pentecostal," and "charismatic," among others) which puts major emphasis on the events, particle-like phenomena. It is probably going too far to suggest that they are stuck in Newtonian mechanism. I have already mentioned some of their particle-like ideas: creation, miraculous healing, and new birth. We can easily add a significant list: 1) specific ministry slots such as pulpit ministry, pastor, priest, or missionary, 2) special revelation in the Bible, 3) dedication services or rites, 4) specific experiences with God, 5) baptism in the Spirit, and 6) gifts of the Spirit. You can and should fill in for yourself what the broader wave-like conception of each of these might be. The priesthood of all believers, practicing the presence of God with Brother Lawrence among the pots and pans of the kitchen, and a call to Christian discipleship for believers in secular careers serve as complements to the chiefs with careers in Christian ministry. *Revelation* does not occur only through the Word, but also through the world; mankind's fallible *discovery* of truth is not only in the world but also in the Word. A single outstanding experience of God's grace will not provide the daily manna; manna stored spoils. Grace must be quotidian, not an occasional feast. Baptism in the Spirit is no substitute for walking in the Spirit. Fruits of the Spirit take time; they are not specific gifts for a specific time only. Love, joy, peace, etc., are lasting qualities of life, fruit of an ever-bearing tree. It may be stretching the analogy to consider all these broader experiences of the Christian life as more wave-like, but the analogy can be helpful.

The liturgical church lists its *particulate* phenomena in the sacraments: baptism, confirmation, Eucharist, communion, confession/penance, matrimony, unction/anointing/last rites, and ordination. Sacramentals, what I call second-string sacraments, such as the sign of the cross, would add to the list. By their definition and in their practice, they function similarly to the list above. As an outward and visible sign of an inner invisible and lasting reality, they represent a particle manifestation, a temporal-spatial event, which signifies a wave-like eternal covenant relationship, not limited by time and space. At least the Eucharist and confession are available on a regular basis in some churches.

These dualities are not mutually exclusive; they have no real contradictions between them. They surely only qualify as rather trivial truth, which we unfortunately have not always mastered. When we come to attributes of God, to his action in the world, to his relationship to those made in his image, complementarity takes on a deeper level, one which may qualify as Bohr's *deep truth* or Kierkegaard's *paradox*.

In physics, Bohr would not have used the word *paradox*, feeling that Nature somehow had a mechanism for avoiding contradiction. Bohr would not have used his *complementarity foundation of all knowledge* for theology, but he did venture into such topics without invoking the name of God.[18] Love and justice were complementary pairs for him, almost certainly seen from his perspective as a parent. Feelings of love did not always work toward justice, and strict justice sometimes left no room for love. Yet of course he knew the need for both sides. Development of character and responsibility must have felt contradictory to the need for safety and avoidance of risk. Witnessing his own son swept overboard and lost at sea must have seared that into his consciousness. Even though the mysteries of theology and religious life would seem to be a gold-mine for such conceptual probing, we do not find any such discussions. Margrethe, his wife, said he wanted to write a book on religion.[19] We do not have even a decent start or excerpt. I think he must have known that such was beyond his depth, as it is for most of us. Always leery of venturing into anything approaching a realm of the spirit, he nevertheless, in private correspondence, noted that the problems that scientists confronted in this world of atoms, beyond our sight, was not so different from that of the prophets trying "to describe the nature of God on the basis of our human concepts."[20] The foremost authority on the unseen world of the atom at least saw the comparison with another unseen realm. A leap of faith, an acceptance of new paradoxes, and a willingness to put decades of hard work into the harder questions which religion faces might have made a big difference in his understanding of the world of God.

If we wish to make sure our God is logical, rational, and lacking in contradiction, we may also shy away from the strong word *paradox*. If we have faith in God, as unsafe and unsanitary as he may be, we may at least be as bold as theoretical physicists and declare the paradoxical nature of God as necessary and positive. I will leave that to you.

Theologians seem to like long words to encode these paradoxical mysteries of God. Let me merely give you a list which I heard from Vernon Grounds, then president of Conservative Baptist Seminary:

1. Ontological paradox: tri-unity in the being of God.
2. Cosmological paradox: God as separate from but completely involved with his creation; transcendence/immanence.

18. Bohr, *Atomic Physics and Human Knowledge*, 152.
19. *AIP*, Oral interviews.
20. Petruccioli and McGilvray, *Atoms, Metaphors, and Paradoxes*, 26.

3. Epistemological paradox: knowledge of God coming through revelation.
4. Anthropological paradox: man as free yet predestined.
5. Christological paradox: divine/human nature of Christ.
6. Soteriological paradox: salvation showing God's mercy/judgment.
7. Eschatological paradox: limitless love/eternal punishment.
8. Genealogical paradox: origin in natural processes/origins in God.

In various forms, these (and other) sticky issues have occupied the church's attention for its whole existence. The teachings of Christ: the blessed poor, the first/last, the loving of enemies, the leader as servant, the blessing of giving, the present but ever coming kingdom, and the victory represented by the death of God on a cruel instrument of execution only add to the list of seeming irrationalities. Neither Nature nor Supernature follows simple rules of logic and common sense.

Science has earned its privileged position in the world of knowledge by exhibiting its faith in the order, explanatory power, and usefulness of its activity. QM can speak of two contrasting views of Nature meaningfully by insisting that the observer chooses and sets up the situation to measure the specific attribute of interest. This requires a faith and a commitment. We must always do the same.

To be or not to be, that is the question.

To be *and* not to be, that is the answer.

With apologies to Shakespeare. We have set up science and faith as exclusive alternatives when they naturally are and must be complementary. We have many pieces within the Christian faith which are worse than awkward to those of us bitten by the certainty, clarity, correctness bug, sure we have to get it *right* or we will be laughed at by the scientists who think they have it certain, clear and correct. When we fear letting God, the Eternal Being, be tainted by the dirty world of becoming which we live in, we try to shut out the God of music, of history, of becoming. The Christian message, if I'm reading it right, is that God was not *above* coming into space and time. "Because eternity was closeted in time, he is my open door to forever."[21] We must not be ashamed of letting our God get into the closet, or we shut our entry door into the mysterious.

21. L'Engle and Shaw, *Wintersong*, 79, a Luci Shaw poem.

14

Rebellion in the Ranks

> Chance, limbo realities, and dual natures proved too much for Einstein, de Broglie, and Schrödinger, who were clearly yearning for a more complete, more visualizable, more determined world like classical physics.

JESUS'S TEACHINGS WERE ABOUT a new Kingdom, the Kingdom of God, a Kingdom of the Spirit, a Kingdom of the unseen. Read the Sermon on the Mount and you see a real visible rational world on the surface, always followed by a "but . . ." and a new Kingdom viewpoint. In the visible world of laws and praying on the street-corner and doing good by loving your neighbor, you get your reward, which, by the way, will rust and fade away; in the Kingdom of God, you step into a new world of living in the Spirit well past the letter of the law, a world of loving when it doesn't seem to offer any reward, of doing good deeds and fasting and prayer in secret, a world in which God's will can sneak into the earthly kingdom. Particles and lawfulness will not go away, but they are undergirded and guided by an unseen wave, a spirit which tells them and us where to go, not as a law but as a choice. If I read the church correctly, we still haven't come to grips with this revolutionary world. And I guess the big take-away point for me is that we will continue to struggle with this until we get a new body in that new Kingdom.

Bohr of course was not exactly the Messiah and they did not crucify him. But he did catch sight of some revolutionary ideas that got some tremendous push-back. Mysterious wave/particles that *require* our active choice to be manifest may be real (or not), but they don't fit the gospel according to Newton's followers. A physical world having a mind of its own somehow integrated with a world of conscious mind was too much, yes, even for Bohr. Bohr would get close, even uttering the words *irrationality* and *mystery*, but he could really only integrate the quantum world with a classical *physics* world, not a world of mind, let alone spirit. The reason he felt the sting of Einstein's rejection of the quantum world was that he also possessed the flesh and blood of a classical realist. Some have cast Einstein as the classical theist (with Nature as his God) and Bohr as atheist, willing to live with the consequences of his unbelief (in the Nature of classical physics).[1] I would suggest that Einstein was the classic *de*istic Jew, maybe the Pharisee (with none of the common negativity we have attached to their serious and sincere adherence to the law). Bohr at least took a little step of faith to believe a new kingdom was at hand; maybe he is John the Baptist, rather than the pope (as he was sometimes called). At least he saw a need for a messiah, a new worldview which he could not put together to the satisfaction of a culture two or three centuries deep into the domination of classical physics.

De Broglie tried to maintain a simple deterministic particle motion by treating the waves (which he had invented in the *material* world) as guides for the particles. Although the record is not clear or complete, Pauli, the physics world's Sword of the Lord, apparently cut him off with questions he could not answer, producing a cowed orthodox Copenhagenist for a few decades. Schrödinger wanted to make the electron into a wave, a compressed wave which therefore could act like a very small particle. But very compressed waves, wave packets as we would call them, basically destroy themselves, not as *waves* but as *packets*, by rapid spreading; this was clear old-fashioned wave theory. This was the counterpoint to the classical *collapse* of the atom, a classical *explosion* of any wave too tightly confined. Einstein was determined that Nature could and did beat the uncertainty principle. For him, the shuffling of the deck in a measurement was well-choreographed by the inscrutable Old One, but it was a magic trick, a sleight of hand. The card did not go into some probabilistic limbo awaiting a so-called incompatible measurement. We may not know what the result will be, but the result of a second measurement is an observation of a reality which *was there*, waiting to be revealed.

QM had not abandoned determinism. The equation for ψ, the wave function or state function, followed the Newtonian calculus and was as determined and exact as classical physics was presumed to be. But it only gave

1. Jammer, *Einstein and Religion*, 233.

probabilities, not actually occurring events. This gave us two problems: 1) How does the process of measurement, i.e., getting a *specific single* result, relate to the deterministic evolution of an *unspecified array of possible results*? 2) How do specific results manage to be in probabilistic, even random, limbo and yet manage to make lawful, correlated choices?

Both Einstein and Rutherford had seen the problems of the second question already at the earliest glimpses of Bohr's old quantum theory in 1913. You do not take a step on a ladder without knowing what rung you are going to land on. (I did once, and a bush below limited my bodily damage without suffering fatal damage to itself.) If an atom is a ladder of energy states, starting at an excited state (an upper rung), it must know which among several rungs it is going down to; that is what specifies the frequency of the light emitted. Moreover if the energy lost in the step down is going to appear later at some absorbing detector atom, the emitting atom needs to know which direction to send the energy and which direction to recoil with opposite momentum. Bohr had managed to ignore or delay such questions by not believing in the light *particle*, and, for several years thereafter, not requiring these conservation laws to hold in individual cases. If there is a single particle produced, a single particle received, and if energy and momentum conservation are to apply, the probabilistic limbo state activated into specific results by a measurement process is problematic.

Bohr's "old" quantum theory (1913) says that there are stationary states of an atom and jumps between them correspond to emission or absorption of light. Einstein had since 1905 held that the emission of light was a creation of a light particle which could be detected by annihilating the particle and taking its energy. The light particle thus reasonably, by all the logic we could understand, traveled at a finite velocity c between the creating atom and the annihilating atom. The creation process and the direction of travel appeared, and still appears as far as we know, to occur randomly in time and space, with the probability in a given time period being characteristic of each atomic transition. All directions are equally probable and the probability wave starts at the potentially emitting atom and goes out like a balloon expanding as it is blown up. This process of going in all directions is

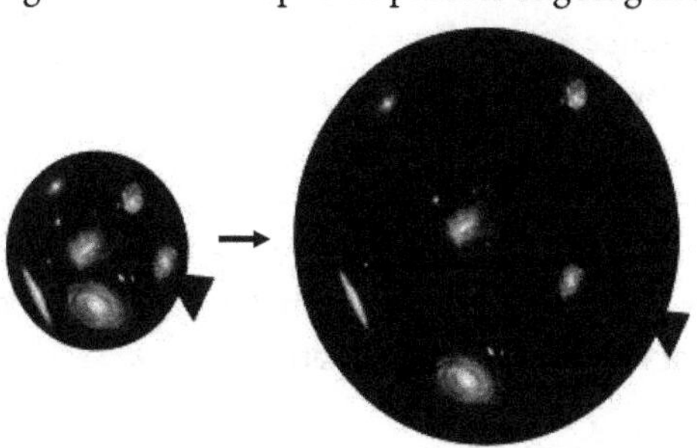

not at all particle-like, representing an essential wave nature, but the beginning and end which we can see by recoil kicks and energy exchanges always act like whole indivisible particles.

Now there is an account of this with which Einstein would, I think, be happy. The quantum mechanical wave function which goes out as an expanding sphere represents our limited knowledge and therefore considerable ignorance of what is going on in an unseen world. The emitting atom holds its lottery, perhaps randomly or perhaps in a real process we are not privy to, deciding when, which of several frequency choices, and in what direction to send a photon. It does so, and the photon goes out along a specific direction, making itself available to be absorbed by any appropriate detecting atom in its path. When a detecting atom wins the lottery at its location, getting all the energy of the photon, our ignorance is cured and all is known or at least knowable. Thus we have an incomplete but purely realistic and local account. We don't and maybe never can know everything (that's why we have to use probability), but everything is above-board, no magic, no action at a distance.

But QM theory as it developed from real experimental results did not talk that way. The electron or the photon in a two-slit diffraction pattern did not seem to go through *one* of the slits as particles should. When we checked to see which slit, we destroyed the pattern which we understood as being caused by something going through both. The wave function which we wanted to interpret as knowledge seemed to have much closer connection to a reality. *Not* knowing or *not* being able to check which path seemed to mean that it in fact did not choose but took all paths. Thus developed an interpretation which may be called "popping of the qwiff" (quantum wave function). Between the measuring points, ψ develops according to a wave equation; in this case, like an expanding sphere, the balloon. At the detection point, the balloon is pricked; the wave collapses, leaving no chance for detection anywhere else. Wheeler describes such descent into ignorance or blindness, punctuated at two ends by clear knowable happenings, as a great smoky dragon whose body is covered in smoke but whose head and tail make themselves clearly known by bites and stings. The wave function was not just a measure of our ignorance, but the reality of traveling on all the paths.

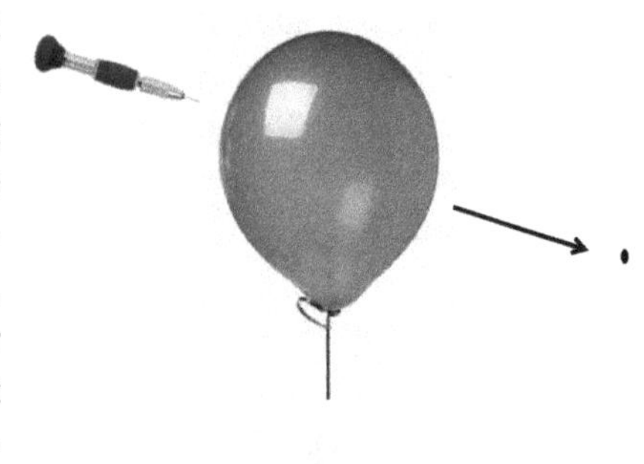

Einstein had the clear sense that this was unacceptable; it required connections which had to be instantaneous over distances in space, breaking his speed limit; it required *instantaneous* collapse, effects traveling at *infinite* speed. Newton had seen the problem with gravitational forces acting at a distance, declaring openly that no one could and should believe in such.[2] Einstein had merely added the time barrier; influences could happen by sending messages, information, or particles but this cannot be done instantaneously. A particle detected was like an *indivisible* lottery winning; the wave was like an extended, traveling network of lotteries which had to be separated to cover all the possible sites but also had to be instantaneously connected to prevent two winnings, two particles, when only one was available. A particle disappeared into a smoky smeared-out unknown existence but always had to come back as a single particle; *one* sting at the tail may apparently divide and take all possible paths in the cloud of unknowing but it must come out the other end as *one* bite.

This was not a clear determinism vs. probabilism clash, nor a clear wave vs. particle duel. There was simply an impossible combination of these aspects. Einstein believed, with good reason, that God made sure that energy and momentum (and many other quantities) had strict conservation laws. How could he let the process get away from him by randomizing any piece of it or letting anyone make an independent choice and still recover the laws, the order, the absolutes, the control? In that sense, Einstein represents a classical theism like that of R. C. Sproul[3] who cannot let his God exhibit or use the smallest iota of chance. The fact that our choices are redeemable, that choice can be and is trumped by Choice, God's Grace, is the gospel of Christianity. Hard, maybe impossible from our perspective, but nevertheless real and true in God's world. Mark Twain, not known for his Christian faith, nevertheless hit it right on the head: "Heaven goes by favor [*grace*]. If it went by merit [*works*] your dog would go in and you would stay out."[4] (Bracketed words are the traditional Christian language translations

2. Letter of I. Newton to R. Bentley, 25 Feb 1693. Quoted in Jeans, *Physics and Philosophy*, 116.

3. Sproul, *Not a Chance*.

4. Paine, *Mark Twain, Vol. 4*, 1567.

added.) God turned the world loose but he has the End in his hands. Our lack of knowledge is clear and certain, but there are laws and relationships which will always work out. God's knowledge or lack thereof is uncertain to me, but I believe in certainties in the midst of any and all uncertainty. What those certainties are may remain uncertain.

Probability as quantifiable ignorance is acceptable and understandable, but a real unavoidable built-in probability cannot travel, allowing separation and yet remaining connected to produce lawful relationships. We call this connected separation *nonlocality*. If you believe in Einstein's speed limit c, you don't believe in nonlocality, a manifestation of an instantaneous, continuous, unavoidable wholeness of the universe. "*Raffiniert ist der Herrgott, aber böse ist er nicht.*" Subtle is the Lord God, but he is not malicious. (Einstein) You might have guessed that *raffiniert* (translated here as subtle) is a cognate for *refined* and you would be right. But ask any German and I think you will find a notch of negativity in the word: artful, crafty, cunning, sneaky, at least playfully deceptive. For Einstein, it would be evil of God to play such nonlocal tricks and even an impersonal God or *especially* an impersonal God would not do that. Seeing in the dark is tough, in science and in religion, and God seems to give us all at least some difficult patches of darkness.

Schrödinger gave this impossible nonlocal connection a name, *entanglement* in English or *Verschränkung* in German.[5] When two things encounter one another, or are produced in a common origin, they are entangled in such a way that spatial separation does not keep them from that original bond. That bond is usually expressed in conservation laws. If their encounter had a certain amount of energy, momentum, spin, or other conserved quantity, the many paths which might seem to separate them and allow indeterminate amounts of various quantities does not allow

5. Wheeler and Zurek, *Quantum Theory*, 162.

violation of the conservation laws. It is difficult and perhaps dangerous to try to read subtle connotations of words, especially across language barriers, but *entanglement* seems to connote a mess, not an orderly connection. That's the bad thing that happens to fishing lines. Schrödinger was certainly an opponent of the conventional reading of QM and may have wanted to denigrate a particular understanding by such negative connotations. In any case, the name stuck, just as Hoyle's mocking "Big Bang" designation for an exploding creation of the universe became standard terminology. The German word seems to have a variety of nuances: the verb *verschränken* means to fold, entwine, interlace; the noun *Schränke* means barrier, limit, boundary; the noun *Schrank* means a cupboard or closet. The world is bound up, interlaced in ways that we cannot imagine. No mechanism within our ken, our imagination, or our faith seems to account for it.

As a radioactive atom awaits its disintegration, a transition to something different while producing a ray, our realistic thinking sees an impending event with growing probability, an unfulfilled expectation until its time comes. A consistent QM picture contains an evolution of a superposition, a combined reality of a "mother" state and a state with a "daughter" and the ray. In a time which we call the half-life, the state will go to a state which is half-mother and half-daughter/ray. Just as we speak of a state of an electron going through both the left and right slit, we mean here a state which is *both* an original state *and* an altered state including a ray which will fly off with lots of energy. For the ray, this is really a state of *both* "not to be" *and* "to be," a state of *not being* and *being*. This sounds ridiculous, but it is in a world we cannot inhabit, so we can perhaps tolerate such fuzzy language. After all we cannot prove them wrong. Schrödinger saw that such a ridiculous picture could be brought up into our world if QM is a general law, not just for the atomic world. Put this ridiculous atom in a sealed steel box where we have enforced our lack of knowledge or ability to see and measure. Also place in the box the unfortunate proverbial Schrödinger cat and a Geiger counter attached to a mechanism which breaks a glass container of poisonous gas. One half-life later, the sealed box contains now a half-live, half-dead cat. It is not *dead or alive* as our realistic world imagines it; QM says it is dead *and* alive, in equal portions, and I don't mean various parts of the cat. This is Schrödinger's famous last-ditch memorable volley at QM, described in a single paragraph of a 1935 paper, clearly showing how ridiculous QM was.[6] Superposition, the ability of a system to be in multiple states at once, was simply ridiculous. Cats are simply dead or alive. They may have nine lives but they don't therefore have eighteen half-lives.

6. Wheeler and Zurek, *Quantum Theory*, 157.

Toothpaste For Dinner.com

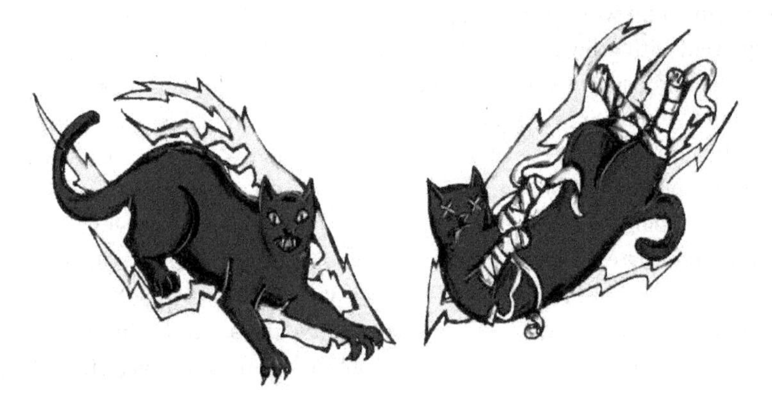

Cartoonist's rendition of a superposed half-alive, half-dead cat

Schrödinger was arrested for cruelty to animals. His fate is uncertain.

Einstein saw the entanglement of two particles in a single creation as an opportunity to beat the uncertainty principle or to show the incompleteness of QM, to prove a reality which QM did not know. In particular, he posited *twin* particles. Heisenberg's uncertainty principle had said that a measurement of one variable unavoidably affected a subsequent measure of its incompatible partner variable. If we separate twins, we know that measuring on one twin cannot be affecting the other; that's what separation means and we can do this in ways that will be enforced by the speed limit of the universe. We measure the position of one and "measure" its momentum by measuring its twin's momentum, knowing they are equal but opposite, having separated from the common origin. The reality which Bohr denied, the *specific* momentum when its position had just been measured, must be there by virtue of a distant measurement which could not have any effect. Einstein said the momentum of a twin was *real* if it was predictable with certainty. Bohr agreed that it was predictable with certainty. That is, Einstein's measurement on the distant twin would always allow certain prediction of the momentum *if* it was measured. But here QM can split hairs in ways you will find hard to swallow. *If* it is measured we know what it will be. If it is not measured, we don't know what it is. And since it is not measured *if it is not measured*, you cannot prove them wrong. Bishop Berkeley had said, "To be is to be perceived"; Wheeler, a disciple of Bohr and believer in the Copenhagen interpretation, said that it *is not* unless it is observed. Bertrand Russell whined in his later years that he and his philosopher friends of earlier years could scoff at Berkeley's idealism, the reality *created* by the mind, but now, after QM, they were all taking it as gospel truth.[7] A counterfactual statement, "If X, then Y," doesn't tell you much of anything when X does not apply. "When pigs fly" and "when hell freezes over" are more effective protections for liars and promise-breakers than merely crossing your fingers behind your back. The classical physics ontology allowed us to pretend we were pure spectators (subjects) nicely isolated from an objective world (objects). Bohr liked to emphasize that we are actors as well as spectators. The extreme view is that our spectating *created* the event; at least the event is not there unless we are there to observe. The late arrival of humans on the scene of the universe's history then begs for an explanation of how the world got along before we arrived. Berkeley's answer is obvious: God. And I guess that really was his whole point. Reality created in *our* minds did not point to us but to God and our relationship as image-bearers and co-creators with him.[8] A reality out there independent of us pointed to a reality out there

7. Russell, *Autobiography, Vol.* 2, 292.

8. Polkinghorne, *Quantum World,* 66. Ronnie Knox wrote a limerick about

independent of God. At best this left us with a deistic God who abandoned his creation. Even if he left it with enough law and order to satisfy Einstein's need for awe and wonder, this kind of God doesn't care enough to interact with his world.

Einstein also published this, his last grand attack on QM in 1935, with Podolski and Rosen. The paper was like a thunderbolt to Bohr, but the rest of the physics world let Bohr worry about it. I do not believe a single professor told me about it (my education was near 1960). It has become justly famous since then so that it is now recognizable by the abbreviation EPR for *E*instein, *P*odolski, and *R*osen. EPR's treatment of the traditional position-momentum pair of Heisenberg's uncertainty principle was expanded by Bohm using spins (or polarizations) as the entangled variable. This eventually not only captured the imagination of the physics community but also made the original unrealizable Gedankenexperiment (a thought experiment, intended for the mind) into a laboratory experiment. As a result EPRB became widely used as an abbreviation for EPR *B*ohm. Bohr's answer to EPR (published in 1939) is not very understandable or satisfying to those outside his frame of mind. He stuck by QM and the strange entanglement and superposition, in spite of its strangeness to a realistic mind. The set of two particles must be treated as an inseparable system although we can see no mechanism for what appear as mysterious, magical, or even impossible connections with apparent separations.

Already in the early 1930s, with discoveries of the neutron, of antiparticles, and cosmic rays and Pauli's suggestion of the neutrino, the world of physics was aiming toward a world of nuclear physics with the all-too-well-known climax of the Manhattan Project and two B-29s carrying nuclear bombs over the Pacific Ocean. Hitler's rise in 1933 had begun the separation of good friends Heisenberg and Bohr. Bohr was working on models of the nucleus in the late 1930s and carried the news of the possibility of chain reactions to America. He was sworn to secrecy, which would probably have been impossible for a man accustomed to working out loud in collaboration with people from all nations. His assistant, Rosenfeld, was not sworn to secrecy and had been cooped up with Bohr for the long trip across the sea. The inevitable happened.[9] Heisenberg chose to stay in Germany, ostensibly

---

Berkeley's idealism: There once was a man who said "God / Must think it exceedingly odd / If he finds that this tree / Continues to be / When there's no one about in the Quad." The rejoinder, approved by Bishop Berkeley, by an anonymous author: Dear Sir, Your astonishment's odd, / I am always about in the Quad. / And that's why the tree / Will continue to be / Since observed by yours faithfully, God.

9. Bernstein, *Quantum Leaps*, 58.

to save German science.[10] His place at the head of the German nuclear program and a strange visit to Bohr in occupied Denmark during the war completed an estrangement of this entangled pair.[11]

QM appeared settled and set, however unsettling its mysteries seemed. Heisenberg and Dirac both wrote books before 1930; Dirac's still serves as a standard. In 1930, the famous mathematician John von Neumann took on the task of laying the mathematical foundations. As we all know, mathematical truth is certain and trumps empirical truth. (I probably should not tease, but you know this is widely believed.) Certainly, Hilbert[12] did not think he could trust the physicists with the mathematics. Von Neumann's book[13] was dealing with a strange world and of course could not stick to its knitting (of pure mathematics). It made obvious frontal attacks on two of Einstein's objections, the measurement problem and the question of a hidden reality behind the QM façade.

Von Neumann decided that he must bite the bullet on the measurement problem and admit that although the dynamical equations handled the evolution of the wave function ψ quite correctly when no measurements are going on, they simply don't work when a measurement takes place. The dynamical equations describe the inflating balloon, an expanding ψ, but cannot describe the pin-prick and resulting instantaneous collapse. In spite of his brilliance, this is as easy to caricature as Bishop Berkeley. It is unassailable to say the equations work when no measurements take place. That has no testable empirical content because you cannot check it without a measurement. This is perhaps unfair because the equations of evolution do provide predictions (at least of probabilities or statistical distributions) which can be checked by subsequent measurements and appear inconceivable without the framework of QM. But von Neumann, for all the justifiable hype of his talents and for all his good intentions, leaves the problem of measurement in a very unsatisfactory position. Two laws apply, one for measurements and one for no measurements, but exactly what a measurement is and which law thus applies is not adequately clarified. We have been left with the problem.

I don't want to put down a great mathematician. When I graduated from high school, I could of course name Euclid and Pythagoras as great mathematicians, but von Neumann was probably the only twentieth-century

10. Powers, *Heisenberg's War.*

11. Brennan, *Heisenberg Probably Slept Here,* 174. This incident was dramatized in a play: Frayn, *Copenhagen.*

12. Lindley, *Uncertainty,* 90.

13. Von Neumann, *Mathematische Grundlagen.* In English, *Mathematical Foundations.*

mathematician of note I could name.[14] But von Neumann's treatment of hidden variables, some unknown set of quantities which supposedly specified the real values of our traditional realities that Einstein and most of us believed must be there when QM only gave us possibilities and probabilities, turned out to be simply wrong. He was supposedly putting a *solid and certain* mathematical foundation under the world of physics; he proved a theorem that says hidden variables cannot exist, that the shadow land of unreality must hang over the unseen world, that nobody, not even God, could possibly produce a deterministic world that gave the results of QM; and *he was dead wrong*. A mathematician's nightmare. We'll come to that next.

14. I had attended a summer "math camp" named after von Neumann.

15

Where Do We Go from Here?

> Von Neumann, a legendary mathematician, presumably proved the impossibility of the deterministic classical world to explain quantum phenomena. Bohm produced a deterministic formulation of QM, opening the door to classical realism.

Nature, we would think, does not have to ask or answer that question. It goes where it's told by laws of Nature; it's pushed or pulled by forces over which it has no control. Nature does not have a choice or give a choice, any chance to behave otherwise. That is the world of classical physics and much of the world of classical theology. God and Nature give no quarter, allow no dissent. Ah! Isn't it just perfect? God and Nature have taken the universe by force. QM of course said: No, there is a fundamental chance built into Nature, a choice that goes from the smallest and cannot help but be made manifest all the way to the largest. Some branches of theology/philosophy say: No, God *had* choice, God *made* choice a fundamental part of those made in his image, and that choice, for better or worse, permeated the whole universe by percolation, diffusion, causation, or whatever model you might be able to conceive. And the wonder of it all is that no matter how much you scratch your head in philosophical/theological musing, read your ancient texts with linguistic/exegetical skill, burn the midnight oil of sophisticated laboratory experiments or observation, Nature and God seem to always leave you with a choice; you are not constrained to believe one or the other. As Sartre realized unhappily, you are *condemned to choose.*[1]

1. Sartre, *Being and Nothingness,* Part 4, chapter 1.

Notwithstanding the impossibility theorems of the likes of von Neumann, all things seem to be possible with God.

Imagine a ball carefully placed on a topographic surface with ups and downs. The ball of course feels the gravitational pull to lower ground and has no trouble "deciding" which way to go. If we are graphically or mathematically describing the situation, we have no trouble "reading the green," saying that the force is perpendicular to lines of constant elevation and forces are stronger where the lines are close together (steep slopes). As soon as light was ever represented as a wave, we could alternatively think of the

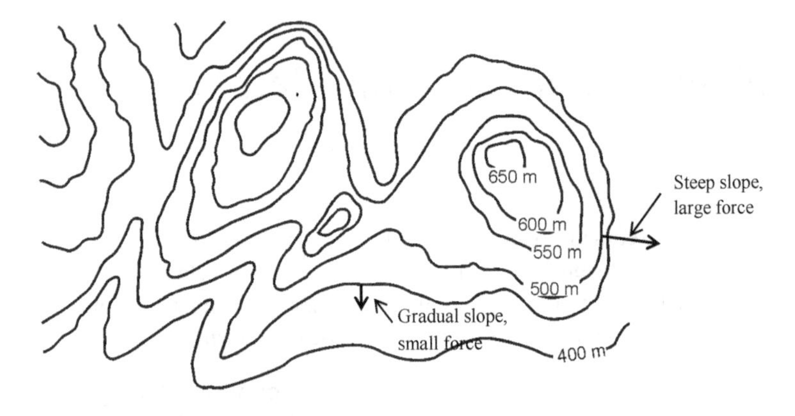

wave as *wave fronts* spreading out from one crest to the next or we could think of the *rays*, the directions perpendicular to the wave front, the direction of movement. These rays are in complete agreement with the ideas of the topographic map above. Either complete set (wave fronts or rays) gave the other picture automatically. Thus the waves and particles of light were historically complementary in almost as simple form as complementary angles: $A + B = 90^0$, where knowing A makes it easy to calculate B. If rays are particles or act like particles or tell particles where to go, a complete picture of the waves gives a complete picture of the particles and vice versa. This may not be easy to carry out in mathematical form, but the concept is clear. How does a particle generate or feel the wave, which effectively tracks all paths and allows all paths to see and feel each other? What *is* the wave? How do the particle/wave manifestations flip into one another? All these kinds of questions may remain, making us feel that we do not have rich enough basic concepts of how Nature works. But, following the historical ideas of optics,

David Bohm[2] was able to transform the mathematics of a *probability* wave into deterministic *particle paths.* He proved von Neumann wrong in the simplest possible way: he *produced* what von Neumann had "proved" was impossible.

No one has been able to prove Bohm wrong. This theory is precisely equivalent to QM, but it guesses or imagines the veiled reality[3] differently. It does not, it cannot, pull the veil away. We still see particles go into one side (e.g., striking a two-slit barrier) and appear in patterns on a viewing screen on the other side. We cannot follow them individually although we can calculate the pattern. QM imagines some kind of pin-ball machine, a shuffling, jostling device whose statistics are calculable, coming out like the results of waves. Bohmian QM imagines a smooth carefully designed green with each particle starting slightly differently (unmeasurably different) and following an exact path to produce the pattern (which is statistical to us because we cannot see and follow the exact paths). Bohm does more, much more than *imagining* this topography. He demonstrates mathematically how it is constructed. Given a carefully constructed input of balls starting on one side, QM gives the probability distribution as it develops in space across to the other side. Bohm tells how to construct the green's topography with traditional forces of classical physics and an additional "force" of the probability wave guiding the particles. *Any* attempt to pinpoint the path by finding the exact starting point changes the green and any change in the green at one point affects the whole green. (Recall the two-slit pattern is destroyed by finding out which slit the particle went through.) The mysteries of measurements affecting the reality, of the apparent ability of a localized being to nonlocalize itself, of the wholeness of nonlocal interactions are still there. But this is your chance, followers of Einstein and über-Calvinists, to assert that chance is a manifestation of *our* ignorance but God or Nature knows which path the particle took, and it had no real choice.

Bohm[4] wrote a textbook on QM right around 1950. That's the way to learn, by trying to teach. His book was fairly standard (still quite good and useful), but shows a deeper questioning and insight than most. Shortly thereafter, he published his reformulation of QM. You would think that Einstein would be elated and that this would level the playing field, taking away the unbalanced home-court Copenhagen advantage. Surely Einstein, not the least bit tarnished in the public's view because of the tilting at QM-wind-

2. Bohm, "A Suggested Interpretation of the Quantum Theory." Reprinted in Wheeler and Zurek, *Quantum Theory,* 369–402.

3. Espagnat liked the concept of *veiled reality.* See, for example, Espagnat, *Veiled Reality.*

4. Bohm, *Quantum Theory.*

mills, clearly holding a commonsense view of the common man's picture of classical reality, and now armed with a realistic, deterministic version of QM, albeit nonlocal, could slay the giant of a dice-playing God, a casino Nature. Surely a sizeable sect of Bohmians would naturally arise. Heisenberg, Bohr, and Pauli were not running scared. Pauli mentions an interaction in which Bohm chides him for not jumping on a potential band-wagon against chance. I do not recall any account of Bohr or Heisenberg trying to confront the new threat; I believe they must have but having read with interest much of their writings of that era, my lack of any recall is significant. Einstein considered Bohm's solution to be "too cheap" and Bohm himself did not count it to be a mortal blow, nor even a seriously damaging blow, against the fundamental intrusion of chance. It deepened his appreciation of the wholeness and unity of Nature and he continued to write interesting and helpful alternative views for decades, certainly upholding the unavoidable chance as part of the natural scheme. There remains a strong, albeit small in number, following of the Bohmian deterministic scheme. Having spent a career in physics teaching, even teaching QM numerous times, I was over sixty before my introduction to the term "Bohmian" in an accusatory "So you're a Bohmian?" It came in response to my questioning some orthodox view, not because of an affirmative credo. This accusatory question had exactly the same feel as "So you're a theistic evolutionist?" aroused by questioning some aspect of creationism in an evangelical church. Theories or articles of faith are underdetermined; they *require* faith. If faith and grace and free will are gifts of God, they require choice. Such choice occurs at the highest conceptual and theoretical levels and it may occur for each and every particle in the universe.

A classical particle or a Bohmian QM particle can theoretically go in any direction and God/Nature chooses one out of the infinite choices. A non-Bohmian QM particle somehow affects all space and can theoretically go anywhere; QM laws and the rest of Nature corral it into probability distributions and God knows (or he doesn't), certainly we don't know, how each individual decision is made. The question of which way it goes is a traditional free-response question and we seem incapable of knowing if there *is* a right answer; our *probability distribution* of answers is confirmed but that's all. The heart and soul of a simpler level of QM arises from systems which cannot handle free-response questions but must be fed multiple-choice questions. The simplest of these is the system which answers True-False, Yes-No, up-down, 0–1 type questions only. The mystery of QM consists in a system's ability to remain in superposition, go through this slit *and* that slit, be both up *and* down, etc., *if you don't ask*, but to accommodate your question with a straightforward unambiguous answer *if you do ask*. It lives comfortably in both an *exclusive or* world and an *inclusive and* world. Coupling the stubbornly limited world of only two answers per question to a world of continuously infinite possibilities is an interesting opportunity.

Which way is up? Even if you don't know, an informed passerby who can answer free-response questions, in this case by pointing in a specific direction, can immediately give you *the* answer. Imagine that your available respondents cannot point and will always give only a *yes* or *no*. Now you can see that the question, the choice of the questioner really matters. If a probability distribution is known, e.g., 100 percent *yes* if you, the questioner, point in the correct direction, 0 percent *yes* if you point downward, and a range in between, you can get a pretty good answer by asking very many bystanders.

Have you played Twenty Questions? A group of people choose an "answer" unknown to the victim who tries to find the open-ended possibility by asking twenty yes-no questions. Twenty questions give twenty bits of information, which is a lot, allowing discernment among a large number of possibilities. (2^{20} is a large number. E.g., only nine yes-no questions are required to find a random integer between 1 and 1000.) And twenty is a good limit to keep your victim from dying too slow a death. It's an OK game.

Dare to try a variation, much harder for the group, and maybe psychologically tougher for the victim. Send your victim out of the room, then decide that the "answer" is *not decided*, but each *yes* or *no* must be truthful for something in the answerer's mind which is also consistent with *all previous* answers. Obviously everyone must take good notes and work hard to keep something in mind; it is probably best to allow players to pass the question to someone else if they are in over their head. You will see that the questions (as well as the answers) play a major part in the determination. One time in our classroom, the "answer" came out to be an obscure nineteenth-century school of philosophy; I was fortunately not asked any of the later questions.

This is how John Wheeler characterized the effect of the observer's choices in QM. We *create* the result, although obviously not completely, by how we question Nature. This of course is not Berkeley's version of idealism, but it is one that shows the power of our questions in creating a reality. As Bohr would have it, we are perhaps unwittingly but unavoidably actors as well as spectators in the drama of Nature. Fred Alan Wolf[5] dramatizes this by declaring that the Universe is the You-niverse; *you* make the Universe. As do I, and all of us.

5. Wolf, *Taking the Quantum Leap*.

Perhaps the simplest conceptual yes-no question of the physical world is the yes-no question of a particular direction answered in two physical systems, by spins of electrons (and other fermions) and by polarizations of photons. If you point in any direction (by applying an inhomogeneous magnetic field), fermions will divide into two groups, *up* (*with* that direction) or *down* (*against* that direction), the two answers being opposite, 180^0 apart. With photons, polarizing beam splitters will divide the beam into two polarizations 90^0 apart. In either case, an infinite continuum of questions (directions) can be asked. Only two questions can have certain answers (a repeat of the immediately preceding question or its opposite, turned 180^0 for fermions or turned 90^0 for photons). All other questions have known odds, but have unknown answers for any individual trial. For any but the two questions with certain answers, we say the polarization is not only unknown, it is doubly unknown, *non-existent*, because we do not even know the next question. If it, we, and/or God know the next question, then it, we, and/or God may go on to have the answer as well. For all who believe in a perfectly determined world, either because of perfect law and order of Nature or of an über-Calvinistic view of God, you are safe; who can deny the possibility or prove you wrong? But it probably won't answer all the questions you or others might have.

Bohm suggested twin spins or twin polarizations would serve especially well for EPR experiments (now known as EPRB experiments). QM claims two measurements at two different angles are incompatible. A measurement of one of the twins in one direction gets a definite answer, now a known polarization; therefore, in another direction, it is in limbo, undetermined if not measured or until measured. Einstein's claim is that each particle *has* the two answers *for real*, one because of *its* measurement and the other because of *its twin's* measurement. Since *any* measurement *could be made* on the twin and would give a definite answer, all of the answers are real, objective physical quantities. Thus the realist position: even if QM does not know the answer (and is thus an incomplete theory), the answer to any and all questions (Is your polarization in *this* direction?) is somehow written in the script or hidden in God's inscrutable mind.

This rather stark disparity between real values, real answers to all possible questions, and the limbo situation of QM's superposition points up the stark choice of our belief systems. A commonsense reality, bolstered by the successes of a calculus-based classical physics, has taught us to view our science and/or our God as the container of all mystery, all secrets. Dumb electrons and photons cannot know or store all the answers. Thus they can live their lives shrouded in mystery; Nature or God takes care of them. We may be much smarter that those dumb small particles, but we also have our limits. We can live hopefully, maybe even comfortably, with that mysterious land beyond us, if we can be assured that Nature/God has control of that land unknown to us. But some of us cannot be comfortable, it seems, to consider that there is anything that is a mystery to God/Nature. Our science and our religions are the exploration of that world of mystery and uncertainty, with some expectations that there is a firm ground down there in Nature or God. Recall the description of the universe as resting on the back of an elephant. But what is the elephant standing on? Another elephant, of course. Below that? Why, it is elephants all the way down. People of faith, in science and religion, are willing to sweep the mystery questions under their respective rugs because Nature/God can handle the tough questions. God/Nature is the place where the buck stops, the place for Archimedes to stand and use his lever to move the Earth. But what if it is mystery all the way down?

Real becoming, the natural result of a real limbo, may be how it appears to us finite creatures, but we are not willing to have that uncertainty extended to God. Limbo, really waiting for the question which is unknown, cannot be part of Nature because it cannot be part of God, our firm foundation where the uncertainty stops. How we think our belief construction protects God from such blasphemy! And it protects God from experiencing time! The Dusty Ones of Israel taught us about a *historical* interaction with God.[6] As William Pollard[7] affirmed, a God who lives with us, who came to us, the Judeo-Christian God, is precisely the God who makes a world of limbo, where time *progresses*, where the future is not bound by the past, precisely the world of QM.

EPRB, the use of separated twins with Bohm's suggestion of spins and polarizations, does not appear to have answered our reality questions. Bohm's deterministic formulation of QM seems to allow us a clear choice: you can believe in a real reality, perhaps known only to God, or you can believe in a limbo reality, brought to fruition by the freely chosen question of the observer. Not only did spins and polarizations eventually give the possibility of experiments, it opened a new window by the consideration of correlations which were well-known by the probabilistic calculations of QM. Can a real deep philosophical question between a deterministic reality and an indeterminate limbo of unknowable or observer created reality yield to real physical experiment? This we turn to now.

6. Cahill, *Gifts of the Jews*.

7. Pollard, *Chance and Providence*, 71, 97,104, 113.

16

Whoa! It's a Dead-end Road!

The hoped-for return to classical realism hits another "impossibility theorem" snag, a theorem that shows that *local* objective reality must give results in conflict with quantum theory.

I LIKE TO TELL my students that we should know the answer before we work a problem. At the lowest, simplest level, this means having a good enough knowledge of the problem that we immediately know when our answer is wrong. For example, a ball struck by a bat simply does not fly forty-two miles, no matter how well you think you have worked the problem. At the highest level, knowing the answer provides a good working definition of theoretical physics. Nature continually presents us with answers, exactly what happened, but we don't know the questions or how we should go about setting up the problem. Producing a model, a scenario, a set of possible questions, is what we are trying to do. The visible answers, readings on a meter, clicks of a detector, etc., are all around us, but we don't know how it happened or what *it* even is, hidden in Nature's many black boxes.

Polarization of photons had a good long history with lots of answers already there for us at the beginning of QM, at the first real acceptance of the existence of photons. In the early nineteenth century, light beams, whatever they were, could be divided into two beams, which we considered *exclusive or* categories, *orthogonal* to each other, to use the geometric language adopted in QM. This classical logic of an excluded middle fit well with the language of particles: a particle goes into this beam or the other beam; there is no third category to fit into, just as a particle was conceived as going through *this* slit or *that* slit. A law of how light divided fit better with a model of light as a wave of some vector field, the electric and magnetic fields developed by Faraday and Maxwell. A vector (a directed arrow of some length) can be made up of two successive perpendicular vectors in what we call Cartesian coordinates. For instance, we cannot usually travel in any direction we like because of fences, rivers, and natural barriers. We build roads and bridges at reasonable intervals. Thus Aunt Ethyl lived two miles west and three miles south of town (following the section lines). You *could* say she lived 3.61 mi at an angle 56.3^0 south of west, but only the crows can do that (or maybe drones). Polarizers or polarizing beam splitters can be turned easily (whereas roads cannot) and we can thus consider breaking beams into two in any number of systems (i.e., at any angle we wish). Only the system where one axis is along the electric vector and one axis is perpendicular to the electric vector results in an *exclusive or* solution, reducing the two-dimensional world to one dimension, 3.61 mi along that particular crow's path. When we realize that intensity or energy or equivalently the number of particles is proportional to the square of the electric field, we have, with results of trigonometry, all we need to know about how the erstwhile indivisible particles must divide, not into fractional particles but by the statistics of choice. The picture shows the traditional vector decomposition into Cartesian coordinates. The vector E can be divided into E sin θ through the y-side and E cos θ through the x-side; the intensity I can thus be divided into I $\sin^2\theta$ through the y-side and I $\cos^2\theta$ through the x-side; thus each photon *chooses* with the fraction choosing the y-beam being given by $\sin^2\theta$ and the fraction choosing the x-beam being given by $\cos^2\theta$. The statistical result of QM must follow the intensities of classical light beams, a formula discovered by Malus, an officer in Napoleon's army, more than a hundred years before QM and the photon.[1]

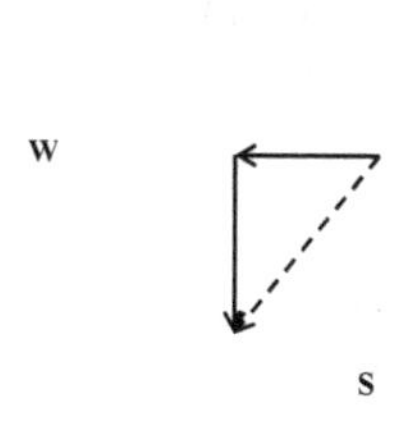

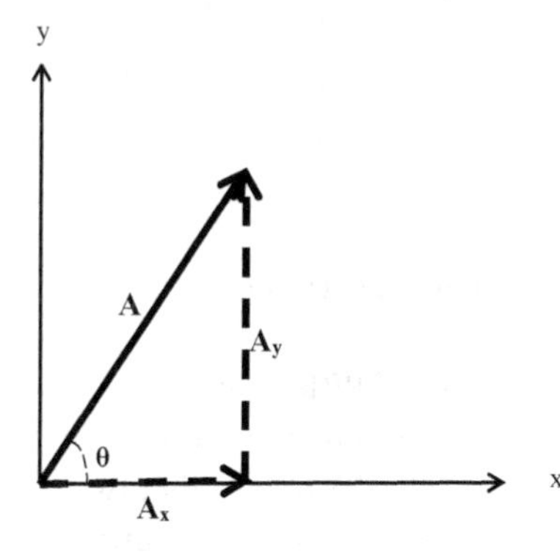

Thus we know the answers. A photon originally prepared by its passage through a polarizing filter will have $\cos^2\theta$ chance of going through

1. Hecht, *Optics*, 278.

the next filter, turned by θ with respect to the first. If you are following the Einstein realism, each photon will be predestined to have the *real* property to go through or not go through any filter you may put up. Therefore, we would say a collection of the photons called o-photons must have a fraction $\cos^2\theta$ which could also be called θ-photons. Every o-photon has a designation (built into the laws of physics or maybe known only in the mind of God) for every θ you can test. We, as theoretical physicists, need only to solve the problem: write any script or prescription which God could know in advance or which could be calculated to specify something consistent with the well-known answer. The problem is a constructive one: make a drawer-full of o-photons, each with some specified value for any and all θs, so that random drawings of large numbers of them in subsequent checks with a θ-polarizing filter at any and all chosen θs will give the distribution known classically.

Let's review. A photon always come in one whole being; it's all or nothing. Moreover, there are two types, polarization in a given direction specified by a polarizing filter or the opposite polarization, 90^0 with respect to the first. It's the real deal of quantum systems: rung one on the existence ladder (NOT THERE) or rung two on the existence ladder (IS THERE), rung one of the polarization ladder (THIS DIRECTION) or rung two on the polarization ladder (THAT DIRECTION). But these things are real enough (carry energy and momentum) that we can set up a situation in which we know we

have one (top rung on the existence ladder). A classic EPRB arrangement is thus as follows: we know how to generate twins and we find the existence of one twin and check its polarization, inferring the *existence* of the other twin and its *matching* polarization. We can check that *ad nauseam*, and it always checks out. But this classic quantum system exists in a classical continuum world; we can turn the polarization filter of the second one to a nearby angle or a distant angle or anything in between, and our photon is limited to *yes* or *no* every time. It makes sense that when you choose a small angle, close to the one determined by the first twin, you will *almost* always get *yes*; turned near 90^0, you will *almost* always get *no*. We experience a *statistical* distribution because the internal workings of the photons or the external determination by God/Nature are simply unknown to us. The statistical distribution is what is given to us by the long known classical results of light as a wave, known by experiments with light intensity and understood by the classical wave theory.

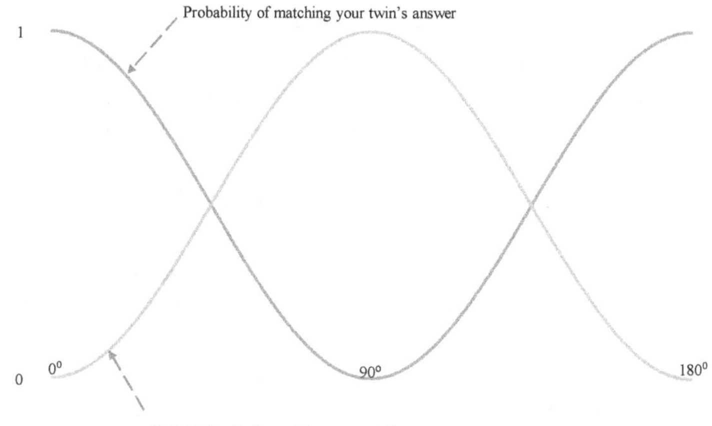

Your task, if you wish to accept this mission, is to put together photon pairs which have a formula or recipe which tells them (by their own inner workings or by a law-giving God/Nature) the answer for each question which can be asked, *before* the question is asked on either end. Each particle starts out with all the answers for any question possible for either of the twins, and the answers must result in the statistics given above. If God knows the future, the questions which will be asked and regulates everything accordingly, he can do it. That means there is no choice *anywhere* in the world. Just as Einstein in science and Sproul in religion would not allow any sliver of chance that would destroy their whole picture, we can believe in such a fixed machine of a world. That of course means that *our belief* is meaningless, because it is simply fixed by the totally determined world we experience. Could God have *had* any choice in making such a world? We certainly could not know because such a world exhibits no choice. If he chooses to make a choiceless world, then he effectively cuts himself off from it. We cannot be in a choiceless world and know a God of choice. This was Einstein's quandary. He believed in a deterministic world, but he knew he could not *act* as though he himself were a deterministic machine. If God were to be anything other than the equivalent of the deterministic universe and have personal characteristics, like choice, then he, Einstein, who experienced personhood in spite of his beliefs, would be responsible to the personal God.

Let's be simple and blunt about it. I, or God, or Nature, or the development of scientific knowledge, have led you, the not-so-street-smart visitor to the city into a dark alley in which you should feel threatened, *if* you insist on believing in the commonsense realism that makes such a model possible. The task I set for you is simply impossible and it is simple enough for you to prove it for yourself. A local realism, a realism (real answers to all potential questions) that does not reach out into the future or cannot communicate instantaneously across space will *not* give the answers of QM, which were the answers of classical wave theory and experiments done a century before QM.

You deserve to see such a proof. If you are unwilling to look at such a proof, as Galileo's detractors refused to look in his telescope, you may of course believe anything you wish. As Einstein said, he had earned the right to be wrong, to make his mistakes. John Bell, an Irish accelerator physicist who spent most of his career at CERN in Switzerland, dabbled in these foundational questions of QM, his hobby. In 1963–64, he proved a theorem which may turn out to be the most important theorem in twentieth-century physics. It is about local realistic systems, assumed to be used in the EPRB experiment, where measurements in each arm *cannot* affect or *be* affected by measurements in the other arm (by definition of locality). One can calculate limits of correlations of answers if those answers are there, real, and fixed independent of the other measurement. Bell did these calculations in full sympathy with Einstein, and found quite simply that the results of such a system did not agree with QM. *A local realistic account of the results of QM is impossible.* Bell's proof, while tied to the simple system we have been describing, polarization measurements of twin particles sent out in opposite directions, aimed for fairly general results and used complicated notations and sophisticated mathematics, which I will spare you. Much simpler versions can show the point quite adequately.

Probably the best known example was published by David Mermin in *Physics Today* under the title "Is the moon there when nobody looks?" (April, 1985). It is elegant for its simplicity and is straightforwardly hard-hitting. He will not take the credit or blame himself, but he anonymously quotes a colleague as saying that if you are not worried about the weirdness of QM, you must have rocks in your head. He proposes the "impossible" machine, the impossibility of making a machine which mimics QM. A central "gun" spits out twins in the middle, one to the left, one to the right. Each path has detectors with three settings 1, 2, and 3 (accomplished in real life by polarizations at 0^0, 120^0, and 240^0) and flashes red (R) or green (G) for the two orthogonal polarizations detected. There are only eight possible realistic designations for each photon: two values (R or G) for each of three settings. There are only nine possible settings for the detectors: three for each arm. The required quantum results (i.e., how Nature works) are: 1) each detector gives R half of the time and G half of the time; 2) if the settings are ignored (but varied randomly), the results of the two arms agree half of the time and disagree half of the time; 3) if the two settings are the same, the results must agree every time. Since the settings will be the same (1/3) of the time, settings which are different must produce agreement of results only (1/4) of the time. [(1/3)(1) + (2/3)(1/4) = (1/2)] This value, (1/4), is the result you can read from our previous graph or from your calculator, $\cos^2 120^0 = \cos^2 240^0 = (1/4)$. You can write all the possible designated answers and you *cannot* reproduce the results of QM with any combination of them. If you are clever, you can probably show that it is impossible. (Look up the article if you would like to "cheat.")

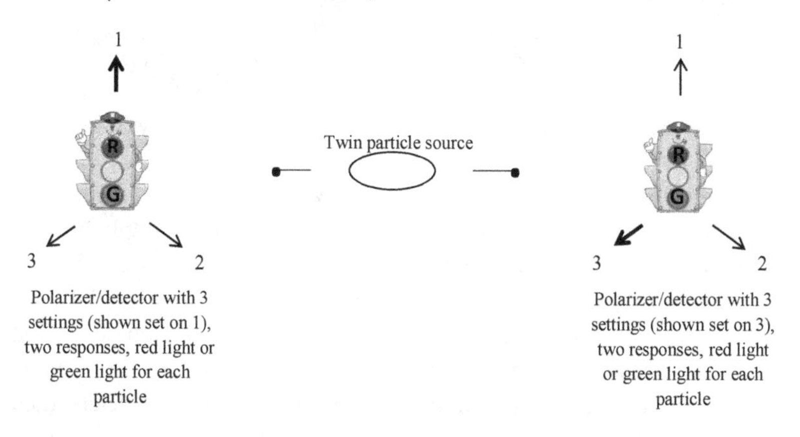

Even though I can no longer write down all the possibilities as we can with Mermin's example, I prefer an example that has many more choices of polarizations and yet easily allows a numerical calculation.[2] Imagine polarization settings at each of the sixty minute markings around a clock face. If the settings in the two arms are the same (or thirty minutes = 180^0 apart), the results must be the same, both *yes* (YY) or both *no* (NN). If the settings in the two arms are different by 90^0 (fifteen minutes or forty-five minutes),

2. Faries, "Personal God," 18–9. This example is inspired by chapter 1 of Bruce, *Schrödinger's Rabbits.*

the results must be different, (YN) or (NY). Now I imagine the thirty *yes* spots to be ships (Y = yacht?) and thirty *no* spots to be blue water (N = cyaN?). Now suppose my goal is to find a boundary, an unprotected flank of the flotilla, by choosing two adjacent settings which get a (YN) or a (NY). Even though I cannot possibly write down the 32,000+ combinations, I can easily visualize the flotilla's best strategy for minimizing its boundaries (while still meeting the criteria of "same for ± 180^{0}" and "different for ± 90^{0}"): lump two groups of fifteen ships on opposite sides (see figure, showing small ships around the edge). With their best strategy, there are four

Best strategy (4 boundaries)　　Poor strategy (many boundaries)

boundaries for my sixty choices; therefore my chances are no worse than 1 in 15. You probably cannot read the QM answer from the graph, but any physicist worth her salt could not only tell you the formula for my chances ($\sin^2 \theta$) but that the numerical answer for our case (one minute on the clock = $6^{0} \approx 0.1$ radian) is about 0.01 or 1 in 100. [Try asking this question next time you meet a physicist at a cocktail party.] Our local realistic model (which does not require determinism by any law but could be determined by the roll of dice or by God's independent choice) gives an answer *greater than* 1 in 15 while God made a world in which the answer is about 1 in 100. As Anatole France said, "If God plays dice, I'll bet he wins."

The wonderful advantage of this look at the problem is that, even though I cannot write all the possibilities, I can scale it up to any number of divisions and get the answer as easily. 1000 divisions in the circle suggests grouping the flotilla into two groups of 250 and chances (with a local realistic world) are at worst 1 in 250. This means that where the QM results are beautifully smooth near zero the local realistic world (LR) requires a kink.[3]

3. Bell, *Speakable and Unspeakable*, 85.

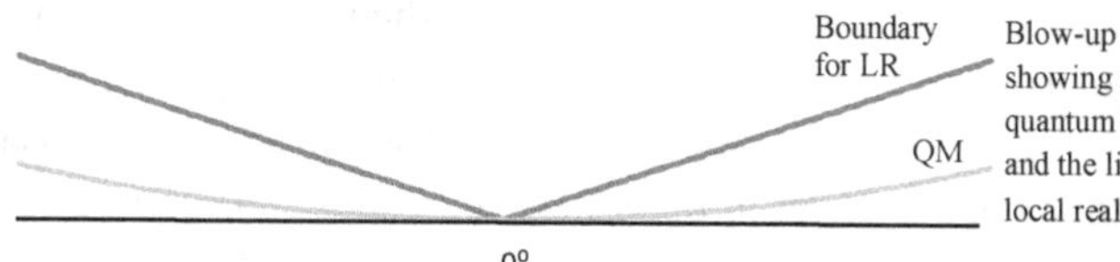

In fact, if one goes to large numbers of ships (thereby approximating a continuum) and supposes they lump together in the optimal fashion, the local realistic world operating at "best strategy" suggests straight lines all the way (shown below). The dashed lines represent a limit for local realistic models, a limit which QM does not obey.

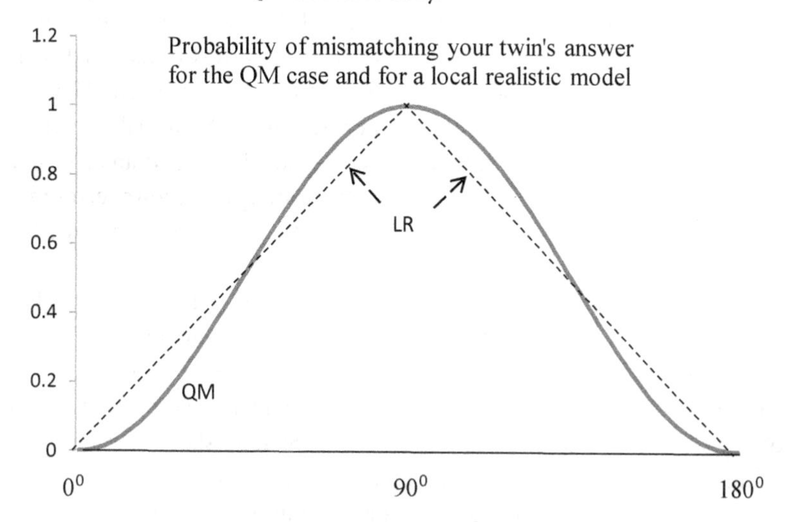

The bottom line is that local realistic models *do not* give the results of QM. The values of the answers to the whole possible range of questions *cannot* be provided or known in advance of the measurement, not even in the deep recesses of God's mind or hidden in some formula known only to the photon or decided in advance by the photon's nerve center or random number generator. It is no wonder that QM would assume a limbo state determined by some chance process *upon* measurement, meeting a probability distribution requirement but not able to be predetermined for any single event. Even though this is a simple version of Bell's theorem, it catches the spirit, the essentials. You can see why it is generally reckoned that it vindicates Bohr, showing the thrust of Einstein's argument to be skewered and burned with unquenchable fire. Einstein and Bohr were both dead by this time, so we naturally have no record of their conclusions to this long, difficult, if amicable battle.

Another powerful image and form of Bell's theorem comes from the idea of message transcription.[4] If a series of measurements of polarization at 0^0 in one arm of the EPRB twin-photon apparatus is considered a message of 0s and 1s, then measurement in the other arm can be considered a copy. In a perfect world, with an error-free scribe, i.e., the setting in the second arm at exactly 0^0, one gets the perfect copy (correlation of 1), error-free (error rate $E(0) = 0$). If the angle is rather some small but non-zero θ, we should get a non-zero error rate, call it $E(\theta)$. Can we now get the error rate for 2θ? Well, we can get a good look. By symmetry, θ or $-\theta$ (counter-clockwise or clockwise) should be the same. That is, $E(\theta) = E(-\theta)$. And the position of 0^0 is arbitrary. If the two arms are set at an angular difference of 2θ, we can think of our 0^0 as halfway between. The message at 0^0 is thus transcribed with errors at a rate $E(\theta)$ in one arm and $E(-\theta) = E(\theta)$ in the other arm. How do the errors add up? Well, they simply add *except* when both make an error at the same place. An error in both arms does not show up as an error but looks like a correct transcription. Thus at 2θ, we always get a lower error rate than the sum of the two error rates at θ. $E(2\theta) \leq 2\,E(\theta)$. See Appendix 2 for a concrete example.

This result is also in contradiction to the result of QM. For small angles, $E(2\theta) \cong 4\,E(\theta)$, clearly way out of the bounds of the result we just got. Where did we go wrong? We assumed there *exists* a copy at the 0^0 chosen halfway between even though it was not measured, exactly as we did before in all local realistic models. The answers to questions not asked must be there waiting for the question, *we realists think*. Our assumptions again lead to a contradiction, meaning that QM says our assumption is wrong. QM again declares and upholds the limbo state, undetermined and non-existent if not observed. Berkeley's dictum ("to be is to be perceived") and QM sound alike.

At this point, the theoretically minded among us will begin to note that we could divide 2θ in an infinity of other ways and declare that $E(2\theta) \leq E(1.5\theta) + E(0.5\theta)$, etc. We could even suggest that $E(2\theta) = 2E(\theta) - E^2(\theta)$ and try to find all the mathematical properties including an exact formula. But we are on a fool's errand. QM says we are wrong with one result. Why multiply examples of why we are wrong? Giving three forms of Bell's theorem (Mermin's impossible machine, finding boundaries of ships, and transcriptions) feels like overkill, because each single argument is definitive. Our overkill is to emphasize: local realism is inconsistent with QM. One of them is wrong.

I have stated this (hopefully reasonably consistently) as a battle between local realism and QM as Bohr, Heisenberg, Dirac, and von Neumann

4. Pagels, *Cosmic Code*, 143–46.

saw it, the so-called Copenhagen interpretation. Believe me, by the time of Bell's theorem, mid-1960s, QM (in the Copenhagen form) was taught as the gospel truth (as it still is). The particular results necessary for our present arguments were known and experimentally confirmed *statistically* by classical results of beam intensities with successive polarizers in a single beam early in the nineteenth century. But no one had done the experiments with *individual* photons adding up to produce the statistics, so-called photon-counting experiments with twin photons flying off in opposite directions. A grad-school classmate of mine, Carl Kocher, set up a two-photon source and investigated the polarization relationship, also in the mid-1960s.[5] When he asked me how they would be related, I gave the standard QM answer. Unfortunately, he did not investigate the range of angles possible so he did not have the ready answer to the stand-off between local realism and QM.

You can understand the conflicted feelings of an experimental physicist facing this conundrum. It is a little like the atheist's view of Pascal's wager. A bet on God and eternal life is worth putting a fair amount into, says Pascal, because the stakes are so high, maybe infinite. An atheist of course has decided the odds are too small. The physicist raised on the gospel truth of QM hesitates to put years of hard work in the lab to test this battle between local realism and QM because nobody blinks an eye if you prove what everybody is convinced is right (QM), and you have a negligible chance of winning the big prize, proving that QM is wrong. Even if local realism is the natural default setting of the *hoi polloi*, the trained physicist has been so brainwashed with QM that it is very hard to risk a career on questioning its very foundations. Suffice it to say at this point that QM may be strange, QM may be full of open mysteries, QM may be open to misinterpretations, but it continues to win all experimental tests.

5. Kocher, *Polarization Correlation*.

17

The Future Matters to Nature and God

> The necessary nonlocality produces a connectedness of Nature across space and time, including the future. God's bag of tricks is bigger and more varied than we can conceive.

I CAN SEE HOW you could be confused. What's the difference between von Neumann's impossibility theorem and Bell's impossibility theorem, that we should scoff at a legendary mathematician and praise a relative dilettante as producing the most important theorem of twentieth-century physics? Will it not also just be proved false by the next bright mathematician or clever experimentalist?

Von Neumann supposedly proved that no hidden variables could account for quantum results; no hidden script of the answers for what actually occurred in a quantum system could be written. That is, although we have results, you could not mathematically write the history that got it there. Bohm proved him wrong by converting the probabilistic script of QM into a deterministic script of what led to the actual results. I should have said: "what *could have* led to the actual results." Everyone was in agreement that these deterministic histories were uncheckable, because the wave functions calculated assumed and depended strongly on the fact that the exact paths were not checked; no measurement was set up to find the exact positions. A QM calculation requires you to state the conditions of the experiment, to tell what you are measuring or equivalently how you are set up. You can maybe see why Einstein considered Bohm's solution "too cheap." You have probably heard the stock market commentators telling you *why* the stock market did what it did today. It feels cheap because it is after the fact and it may be after the facts used to explain it. Why didn't they tell us this morning? In Bohm's world, we are getting the God's-eye overview, knowing after the fact what measurement decisions *were* made. Besides, his scheme allowed nonlocal effects. If you allow nonlocal effects, e.g., that what happens in one arm can affect the results of the other arm, you will have no trouble matching any quantum result of the simple systems of the last chapter. In other words, Mermin's impossibility machine is not only possible but easy if you allow a signal back from the detectors, even if it is a simple wave/particle along the same line as the supposed wave/particles sent out from the middle. Make sure you get the results of the *measured* quantities and make up any nonsense you wish for the non-measured ones. You may not believe them yourself but nobody can check; it's over and done and those were not measured.

Thus Bohm showed that all the answers obtained in a QM system could be envisioned as quantities fixed by some determinative plan (perhaps hidden permanently from us). This may seem too cheap because it appears after the fact and it cannot be checked, but it has the strong hint that it in fact costs way too much. To return to the stock market language, it appears that it is insider trading or price-fixing. "After the fact" seems cheap precisely because "before the fact" means we did not have knowledge. If someone is getting and using knowledge ahead of the fact, we believe it is cheating and should be disallowed, illegal. That is what nonlocality feels like, and it felt particularly bad for Einstein whose relativity was based on an unbreakable speed limit for transmissions of signals and matter in our physical world. Nonlocality appears to be absolutely impossible and QM seems to have incorporated it. Gaining the possible reality of completely determined quantities (albeit unknown and unknowable) seems totally hollow if it comes at the cost of such an unfathomable trick. Feeling that God is cheating by making the odds beyond our understanding is one thing; to think that he does it by peeking at the future is quite another kettle of fish. The former seems like an admission on our part that God's engineering feats are beyond us, but the latter seems personal. Again to think of the stock market: the former feels like the quants with their clever computerized trading programs; the latter feels like simple cheating, insider trading.

The genius of Bell's theorem is that it turns the rules that we may think God should abide by in a physical universe into quantitative results that can presumably be checked in the laboratory, and QM, presumably in God's bailiwick, breaks them, producing different results.[1] These results were known

1. Rosenblum and Kuttner, *Quantum Enigma*, 139.

statistically by the classical intensities of light beams divided by polarizing beam splitters. But none of these classic experiments checked the locality/nonlocality question. We took a beam, filtered it (by a polarizing filter) to make it into a single *known* polarization, then checked how this *known* polarization divided itself when asked about a different polarization. There was no opportunity to suspect any possibility of nonlocality, a measurement at one place producing effects elsewhere; the presumed particle (or wave) comes to the first filter, makes a decision locally, goes to the second filter and makes a decision there. Einstein (in EPR) had used the twins to argue that the identity or reality was there, determined, and written by some script before each measurement. Our natural belief is that, before measurement A, there were lots of photons at different polarizations; filter A just chose to let A-photons through. Because no prior identity was in question, nobody can argue. But now we have an A-photon, and we're going to ask: "Are you B or not-B?" QM declares it is in limbo, neither B nor not-B, but both B and not-B, in such a way that it will be decided with odds that are known. Einstein says it was not in limbo; it *had* that measured identity, e.g., *yes* to A, *yes* to B, built in as a real quality, known by some hidden rule or variable, but just not known to QM (or of course to us). He could "prove" that by measuring A on one twin and B on the other twin. Because A measured with one twin always agreed with the other twin (if and when A was measured) and similarly for B, and these measurements could be (in principle) separated so that no signal could pass between them, Nature had a reality going on that QM did not know. QM was just copping out and calling it limbo and calling it probability when in fact it was fixed all along. Bell had found a way to call QM on the carpet, at least *in principle*. If the system was in fact remaining in limbo until some lottery-like chance operation produced a choice, then it had to have a way of sending effects from one arm to the other even if this violated Einstein's speed limit. The correlation of two supposedly random separated events required a connection. The connection could in principle be ruled out if twins could in fact be separated to arbitrary distances and questioned independently at any time. Everybody believed (which does not prove anything) that the EPRB situation with twins would give the same well-known results as the simple sequence of measurements on a single photon. But light travels so fast and we are so slow in making decisions or implementing them that a simple experiment of EPRB, getting the expected QM result, would prove nothing. The separation *in principle* could not be easily actualized (*in practice*). A typical laboratory-sized experiment has separations of maybe 10 feet; light travels that distance in about 10 nanoseconds (billionths-of-a-second). The set-up time for a measurement is typically hours or even months. The need for action at a distance is laughable;

everything that happens in one arm is easily knowable in the other arm, even if we do not know how it could be communicated. I do not want to downgrade the heroic measurements of EPRB experiments; people put in great effort with everyone believing they knew the results in advance. Everyone was in fact right; QM was acting according to its long-term probabilistic script, but the experiments had not ruled out that little cheat, that signals *could* have been sent. We had not *really* separated the two events and QM had plenty of opportunity to act on insider knowledge rather than doing the incredible magic trick of nonlocality.

This kind of questioning spawned the idea of delayed-choice experiments. If an Einsteinian reality, a disbelief in limbo states, is taken seriously, let us wait until the crossroads have been passed before we make a decision to check or not check, a decision to make one measurement or another. We will delay our choice of measurement until Nature *must* have made its decision. Perhaps the most striking of such experiments, is potentially set up for us with the decision point already in the distant past. Let's see how, starting with some familiar phenomena.

When you drive across the desert, distant mountains are seen doubly (without any drinking), once where you think they really are (*up* above the horizontal) and once upside-down where you think they are not (*below* the horizontal); they appear reflected in that tempting mirage of a lake. I have literally tried, on one such passage through the desert, to walk out to that non-existent water. If you stayed behind me, you would see me walking on the water without getting my feet wet. What is happening is that light is

taking two paths to your eye, one making real objects appear above the horizontal, and one producing the opposite appearance. The bending of rays in the heated atmosphere above the hot surface is responsible. We see this every day on our highways in apparent puddles on the road.

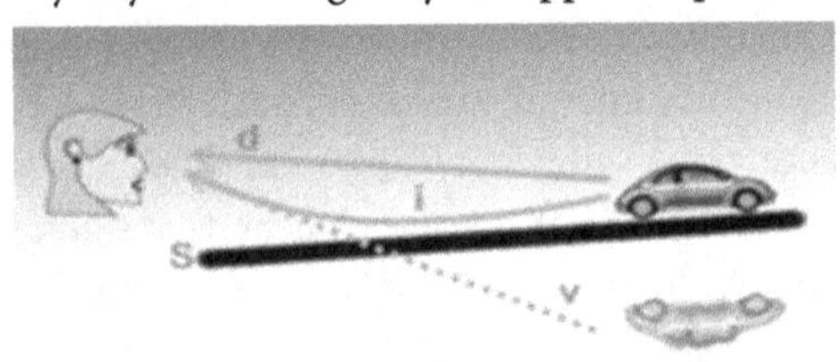

d = direct ray

i = indirect ray bent in hot air

S = hot surface

v = virtual ray, appearing to come from the "reflected" image

Bending of light rays by the gravitation of heavy objects in space provides a very long-line example of this. A star can send light on two sides of a "gravitational lens" and allow us to see faux-twins, seeing the same star in two positions. This light could have been traveling for billions of years. This should remind us of two slits which will produce an interference pattern if we don't ask which path the light took; if we separate the light into that from the virtual star on the top and that from the virtual star on the bottom, we ruin the interference pattern. The light, by our realistic thinking, either went by the top or the bottom path *or* it went both ways and that was all decided eons ago. Yet our decision *now* with the measuring apparatus we set up can distinguish these two events and can effectively act on the "decision" which QM describes as a superposition, a limbo, waiting for the perception to actualize the phenomenon. One needs to be brave or crazy or both to think

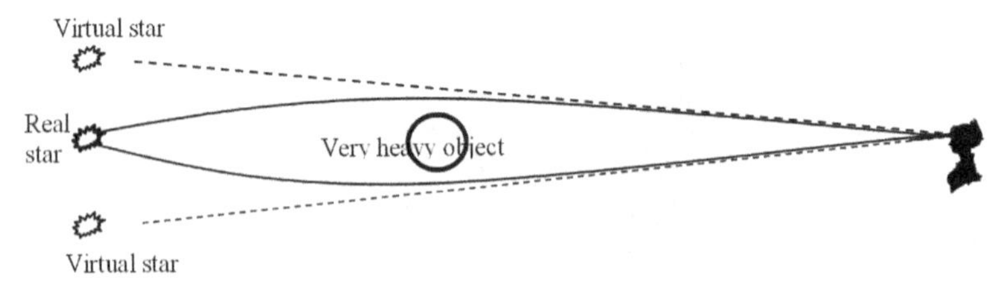

and state that the decision of light from stars sent a billion years ago from a billion light-years away depends on our perception and decisions *today* for its decision at a fork in the road in the distant past. But imagine what that means if God uses the same kind of *natural* processes on some final day to affect our mundane encounters today with two roads diverging in the yellow wood.

Can we bring such crazy experiments into the lab? Yes, if we can make and implement decisions in times like nanoseconds. The best conceptual framework is the Mach-Zehnder interferometer, a fancy "two-slit" arrangement which allows easy changes and adjustments. Instead of two slits we divide the beam, and thus each photon, by a half-silvered mirror (HSM): 50 percent transmission straight ahead and 50 percent reflection at 90^0. By full mirrors (M) we can aim the two beams back to a meeting spot, at which point we can decide to recombine them or to check which way they came. By removing the second recombining half-silvered mirror, we can choose to find if it went the lower path (striking D_2) or the upper path (striking D_1). If

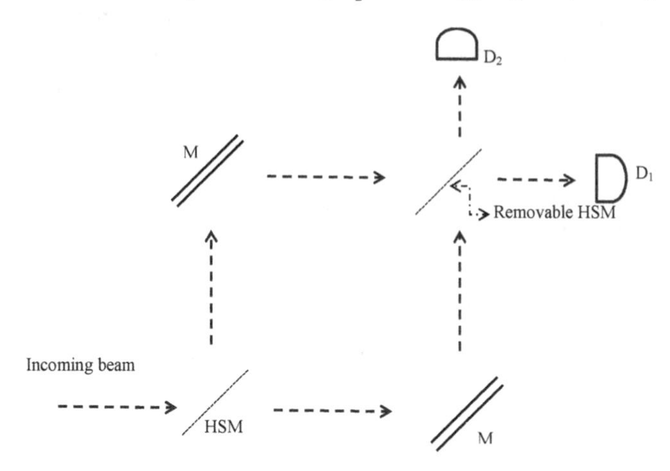

we recombine the beams (using the second half-silvered mirror), then they will show evidence of knowing both paths by interfering appropriately at the two outlets (D_1 and D_2). The numbers arriving at D_1 and D_2 vary according to the interference of the waves which went both ways and could be adjusted by the path lengths. The variation in the numbers at D_1 and D_2 look just like the oscillations seen in the two-slit interference. To insert a second HSM or remove it very quickly is a technical achievement which we do not need to understand, but the bottom line is: you can't fool Mother Nature by forcing her to decide if the photon takes the high road or the low road *before* you decide which way to measure it. QM has its idea of a limbo which waits, answering *Which way?* if that is what you ask but answering *How do the paths compare?* if you ask that. As Yogi Berra would say, "It ain't over till it's over" and until it's over it keeps all options open and lets the earlier realistically presumed decisions reflect what happens later. In an EPRB experiment with two beams, seemingly random events in one arm show evidence (by appropriate correlations) of knowing or being affected by events in the other arm, whether they happened earlier and can be known, or they happened simultaneously and could not possibly have received each other's news, or they will happen in the future. A God whom we try to describe as Timeless Eternal Being, the "I AM" of a tense-less verb of existence, made a world of Nature which was, and is, and is to come. And past tense, present tense, and future tense all matter. He did not make a world to toss aside, making his only connection to it into a *past* creation. He did not make a world in which a causal past completely determines our present and future, a machine made to his specifications and run to his specifications to its fixed end point. He did not even make a world where our past and present interaction with a Creator is all we have. We have a Creator who may look like he has turned us loose to the vagaries of chance or made us cogs in an inexorable machine. We may even feel as though he is with us in the present, pulling for us, helping us. But he is not Lord of all unless he is also Lord of the future. Looking into the future is our ultimate unknown; *future* and *unknown* are pretty good synonyms to us. QM tells us of a basic link between the present and the future. If I'm reading it correctly at all, this is a reflection of a creation that isn't over till it's over, a creation and Creator of the past, present, *and future.* The Jewish tradition gave Western civilization a sense of history but that is not *just* a past. History is going somewhere.

18

OK, So What?

> The nonlocality of QM, the eschatology of theology, and a renewal of Aristotelian final cause all suggest a possibility of a symmetry of causality and teleology. God is not just God of the past and present. He is also God of the future.

> "[T]he idea of teleology has again become possible in the scientific picture of the world." Laurikainen[1]

I BELIEVE IN A directional time; after all, I'm alive which means I have capabilities of using my environment (food, air, water, energy, etc.) to produce some order (my body, ordered stacks of building materials, ordered stacks of paper, ordered strings of words, etc.); besides, my life is all too obviously limited and is heading toward death, the thermodynamic terminus for us all. Directional time and sin are both most obviously empirical; we all experience them. QM reflects both of these well, by strongly affirming real potentials of choice (which can go the wrong way) and the developmental process of actualizations of a specific potentiality which apparently follows or produces a direction in time. Just as balloons do not unpop, popping of the qwiff seems to be irreversible. While life, thermodynamics, and QM seem to provide evidence for directional time in our limited physical experience, I am equally committed to the theological, spiritual, big-picture direction of time which is much harder for individual finite creatures to see empirically. This bigger picture I see by faith, although I can understand if you cannot see it by faith and do not even believe it is there to

1. Laurikainen, *Message of Atoms*, 31.

see by anyone's faith. In spite of my very strong commitment and belief in a directional time, a direction we must go, a direction we may go, and a direction we cannot go, I also see an *un*directed time as an important building block for our world, a time which goes backward as well as forward, in which case no one can tell which is backward and which is forward. If we see movies which include either life (the growth of order at the expense of our surroundings) or decay (friction provides the simple case for mechanics), we know immediately if the movie is running forward or backward. In conservative mechanics, Newton's equations of motion involve time twice, meaning that they work the same way backward and forward. Changing the sign of time is done twice and makes no difference. All Newtonian mechanics is cursed or blessed with this property; you cannot tell the difference in a backward movie of Newtonian mechanics.

Here's where QM does something special, counter to the machine-world of Newtonian mechanics. QM generally makes a biased statement: a measurement actualizes one of the possibilities, making that quality known, measured, and real *after* the measurement, but insists it is not known and not real *before* the measurement. Now I can easily accept that *knowledge* can experience that jump from ignorance to enlightenment; that is often our experience. Maybe it is my prejudice, produced by generations of realistic thinking of *being* as having some permanence about it, but I have trouble believing that there is not a continuity of *being* in that process that we are calling actualization. The measurement is discovering something which was there, surely at least as much as it created something which was not there. I can just feel my (maybe our?) realism percolating up and you may well accuse me of being too much like Einstein, too old to accept the new findings of science, the new view brought about by QM. I will affirm that QM contributes to a real directional time and that the process of measurement and questioning by an observer speaks of our creative abilities, of our co-creatorship with God, and ultimately of God's continuing and future creation of all. But I don't believe that the instant of measurement is the time of creation. Let me explain.

One of the key features we have been attributing to the world of QM is choice/chance. A number of possibilities get narrowed down to one actuality. No knowable determinant appears to be possible. Bohm's theory demonstrates the possibility of an *unknowable* determinant, but Bell's theorem shows that local determinants of answers to *all* the questions we can ask in very simple experiments cannot even *exist* prior to the actualization or determination of the one question which is asked. QM calculates and the world of quantum physics has confirmed that in every case we have been able to measure, the statistical distribution produced by locally determined, really existing answers or properties prior to measurements cannot possibly agree with the quantum world. I hope you can understand why we, physicists, therefore naturally fall into line with a belief in real chance, in real states of unknown and unknowable limbo. The fact that strong conservation laws produce correlated events and that very carefully shaped probability distributions are followed very accurately makes us know that the random elements which seem to be there unavoidably are somehow kept under tight rein. A beautiful and useful theory of how the world works remains even if God is throwing dice. We still can affirm unbreakable relationships, absolute laws, and a statistical causality even when the young Turks are willing to throw reality and causality out the window. Speaking theologically, God can still remain in control while allowing us and *all creation* freedom of choice, even if irrational, mindless, random choices are made. Neither in our theology nor in our quantum physics have we been able to understand how it is done. One of the big lessons to me is the required acceptance, even the bold proclamation, of an unbreakable mystery, the absolute centrality of our unknowing, the inscrutability of God *and* his creation. But I hope you can also appreciate our attempts to penetrate as far as possible into that mystery. It is my experience that a fuller appreciation always comes with our attempts to "scrute" the inscrutable, to solve the mystery, to see into the darkness. The positivistic science of sticking only to what we can sense is a limited dead-end road. We must seek for the things that are not seen. "[F]or the things which are seen are temporal, but the things which are not seen are eternal" (2 Cor 4:18b KJV). Therefore, in spite of the perceived conflict of trying to take the mystery out of God's ways in his creation, I will try. I do not fear, as the late nineteenth-century classical physicists feared, that we will solve all the mystery of the physical world and be turned into technicians and engineers, simply applying our essentially complete knowledge.

So how do I think God can allow chance and limbo, an apparent randomness and loss of causality, and still keep it all tied together and under control? How can nonlocality, the ability of a physical being or event to reach out and touch some other physical being or event without the restrictions of Einstein's speed limit, possibly work in our physical world? Our tendency in classical theology to attribute such absolute powers to God with all the big *omni-* words doesn't help us make the link with the world we live in. And for me, the ability and *insistence* of God's reaching *into* our physical world, not his distance, his impassivity, his inaccessibility, are the key teachings of the Judeo-Christian tradition. I think our solution can be reduced to a single idea which goes by different names in traditional philosophy, theology, and in science. In philosophy, we would call it teleology, the aiming *toward* a *telos* (Greek for "end"), a purpose, a goal, supplementing the traditional causality which we have narrowed into coming *from* a cause. In theology, we might call it eschatology, the concern for the *eschatos* (Greek for "last" or "farthest"), the ultimate destiny or purpose of mankind and the world. In physics, it takes the name of directional symmetry in time, that things, real things or real effects or real signals, go *both* directions in time. Physics was so successful with its very limited view of deterministic causality, a cause preceding and determining an effect that deterministic causality became the standard in both the non-living and the living world. We failed to see that the world of complex systems, thermodynamics, and the world of life seemed to be necessary to provide that direction to which we tied causality. Apart from that directional impulse apparently imposed outside of classical physics, physics could make no sense of a one-directional causality; if time knows no direction, "cause *preceding* effect" is perfectly symmetrical with "effect *preceding* cause," vitiating the meaning of "preceding" and our usual sense of causality.

Somehow the time-symmetry of mechanics, the fact that time always appears twice in the fundamental equations and (-t)(-t) = (+t)(+t) means nobody can tell forward from backward, always just seemed like a curiosity and never hit me with any force. But every physicist eventually sees the strange time-symmetry of electromagnetic waves, and this seemed like evidence which required a verdict, as some Christian apologists like to consider the resurrection of Jesus. Charges which oscillated produced fields which oscillated according to Maxwell's equations, and the oscillations propagated outward as waves, as if to tell the world out there that the charge is wiggling and is willing to share its wiggles with someone. When you write the wave equation which tells how the electromagnetic fields change in time and space, you find that two solutions appear as absolute twins, an outgoing wave and an incoming wave. An outgoing wave sends out its signal, its potential transfer of energy, and the world will get the message at a later time, the delay being the distance divided by the velocity. All seems well with the world; we call this a "retarded" wave with the wiggle at the source "causing" a later "retarded" effect at a distance. The local production of the intermediary field, the meaningful lawful propagation of that field, and the local action of the field on another charge complete something we can accept as real local physical realities. But the incoming wave? That is an equally good *mathematical* solution of the equation, and we call it an "advanced" wave. The effect is in advance of the cause; time is going backward. Physicists usually have a simple way of treating things like this. If you solve a quadratic equation arising in physics and you get an imaginary or complex number or infinity for your answer, you just say the answer is "unphysical" or "impossible" and it goes in the waste-basket. Waves coming back from the future are unphysical; toss them.

Early on, I came to like the strange symmetry of causality and teleology for theological reasons. God was not someone who *only* drove his creation causally from the Big Push First Cause, but he also *equally validly* drew his creation teleologically, eschatologically from the Big Pull Final Cause. The Great Shepherd did not just keep a pack of dogs to drive his sheep in front; he walked in front and asked them to follow him. I'm still fond of that idea and I think QM is at least a whisper of that message from God.[2] He's not going to yell at us or get out the whip, but I think he is speaking.

You can, I think, see that the classical waves have the same basic problem that quantum physics has, the difference being that classical electromagnetic theory has a wave invented as an intermediary between the cause and effect while quantum physics has a particle that cannot be extricated from a similarly invented wave which accompanies it, guides it, or fakes it in some way. If a wiggling charge sends out a reality of a wave in all directions, depending on that independent reality to go out and run into some charged particle and push it to give it some momentum and/or energy, how does the original wiggler know in which way to recoil or how much energy to lose to balance what is happening somewhere else at a later time? This is like trying to balance a checkbook after you have sent out a signed check with the amount left blank. Richard Feynman was working on this classic physics problem with John Wheeler at Princeton University for his PhD when a small matter of the Second World War and the urgency of a bomb project provided an interlude. After the physicists had all become "sons of bitches"

2. Just as Pollard was able to see chance and the asymmetrical time, historical time, created by quantum measurements as the scientific fulfillment of the Judeo-Christian God's action, Huw Price can see that advanced action which results in teleology is the natural result: "With a certain amount of Archimedean gall, we might say that advanced action is just what we should have expected in the world, and that QM simply shows us how God managed to pull it off!" Price, *Time's Arrow,* 194.

by their successful solution of a "sweet problem,"[3] Feynman and Wheeler published their academic work on the classic problem.[4]

What Wheeler and Feynman found for this purely classical problem of electromagnetic interaction was that a self-consistent picture could be produced by giving the advanced wave equal status with the retarded wave rather than chucking the presumably unphysical wave traveling backward in time. What we classicists naturally were seeing as a time-directional set of events ((1) an emitter sends out energy, (2) energy travels out away from the source, and (3) an absorber accepts the energy), was or seemed to be rather a time-symmetrical relationship. The wave was half-retarded and half-advanced and the process of energy going from one to the other was symmetrically an emitter *pushing* it to the absorber and an absorber *pulling* it from the emitter. Because the process of emission *required* the absorber as

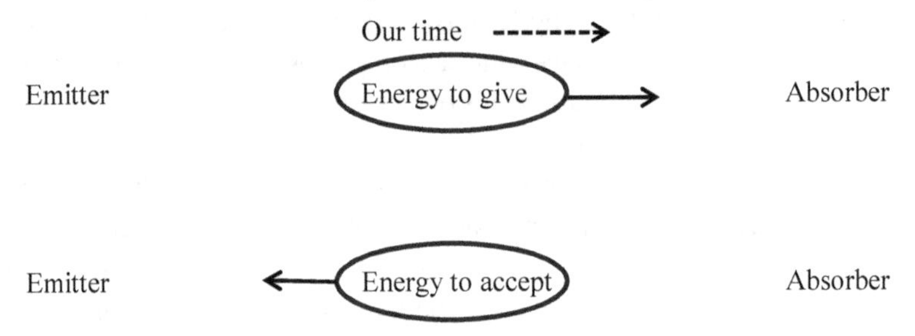

an equal partner, this theory is called "absorber theory." When we say a twinkling star sends out light which I see many years later, this theory says I, the viewer, play an essential symmetrical role in sending an acceptance message to the star: "I will accept your light." Others had dared to make such observer-centered statements.[5] You may note that ancient theories of

3. Physicists had largely ignored the larger more substantive questions of lives, politics, morals, etc., and focused on the many technical issues of wonderfully challenging "sweet" problems, letting the military and political personnel be the responsible "sons of bitches." When they (physicists) successfully built the incredibly destructive weapon, one remarked that "Now we are all sons of bitches." Let me not do the easy thing of condemning them. The "sweet problem" was well-known and would sooner or later be solved; if Germany had solved it first (which was not even close to reality, as we found later), we could have all been dead sons and daughters of bitches; there is a good case for believing that the bomb's deployment saved lives by stopping Japan's continued efforts. Many physicists became sensitized to the unavoidable moral aspects of the presumed independent (of moral issues) basic research and solutions of problems, sweet or bitter. Many, including Wheeler who lost a brother in the war, admirably defended their involvement.

4. Wheeler and Feynman, "Interaction with the Absorber as the Mechanism of Radiation."

5. Tetrode, *Zeitschrift für Physik*, 317.

seeing often had the eye sending out rays. (You can certainly find this suggested in the language of the Bible.) What Feynman and Wheeler did was to give the observer justification as an equal partner in the process.

I think it is worth noting, partly as an exercise in thinking these seemingly unthinkable thoughts of backward time: just as classical second-order equations make -t look no different from +t, a wave or particle traveling backward in space and backward in time looks the same to us as one going forward in space and forward in time. But this means the emitter *gets* a signal back from the absorber at the same time that it *sends* a signal to the absorber. And the absorber *sends* a signal back at the same time it *receives* a signal from the emitter. Everyone, the emitter, the absorber, and a stationary outside observer, all see the same time difference between the processes we call emission and absorption, a time which we call the time of a wave's traversal of the space between. Feynman's thesis works through this classical theory and he gave a seminar on it to the Princeton physics department. Why not make it a quantum theory and see these as creation of a quantum (photon), traveling of a quantum, and annihilation of a quantum, but now a symmetrical mutual action by emitter and absorber with the traveling going both ways in time? I even like to imagine this as a photon which can only be complete by having both its half-photon which goes forward in time from the emitter to the absorber and its partner half-photon going backward in time from the absorber to the emitter. According to the stories, Wheeler was scheduled to present the quantum theory to the department one week after Feynman's talk.[6] It did not happen. Pauli, the famed critic, reportedly was not surprised, suggesting that Wheeler had found it much harder than expected. I am not surprised that I do not know what the problems were and I was too late apparently to get an answer from Wheeler. I wrote him to enquire but he was already quite old and his colleague handling his correspondence did not get me an answer.

Feynman added another piece to the puzzle within a few years by casting QM into a formalism of the time-honored extremum principle.[7] The particle takes all paths, he says, keeping track of phases (like a rolling wheel could keep track of the angular position of a single spot on the rim or a clock with rotating hands can keep track of time), and adds them all up (as vectors).[8] If there is a clear extremum, a path where all nearby paths are essentially the same length, the phases of that path and its nearby equivalents

6. Schweber, *Quantum Electrodynamics,* 382. The story comes from a taped interview with Feynman by Charles Weiner, March, 1966 for the Center for History and Philosophy of Physics, on deposit at the AIP Niels Bohr Library.

7. Feynman and Hibbs, *Quantum Mechanics and Path Integrals.*

8. Taylor, Edwin, *Am. J. Phys.* 66.5 (1998), 369.

are essentially the same and that is the overwhelmingly favored path. (Many vectors are in the same direction.) If nearby paths all vary in length from a given path, the local set of paths together produces cancellation because of varying phases. (Vectors go in all directions.) The particle does not need to be smart and solve any differential equations to know where to go. It tries them all, like any dumb particle can do, if God made it capable of such profligate action. These of course are the kinds of possibilities of a wave-enabled particle. Feynman-Wheeler absorber theory had opened a new wrinkle in this schema: not only could the particle go all paths (a reflection of its wave nature) but it could make each round trip *in no time* by alternately going there forward in time and coming back backward in time. If a single particle can try all the presumably infinite number of paths, why could a single particle not play the part of *all* such particles in the universe? It is no wonder that the imaginative Wheeler could entertain the idea that there is only *one* electron in the universe. If some of us mortals can think of such creative possible forms of the physical universe, how much fun must God have in putting it together, letting us understand *some*, and yet confounding our understanding!

These ideas re-emerged in the 1980s in an interpretation of QM by John Cramer, a nuclear physicist at the University of Washington, an interpretation he called the transactional interpretation.[9] Electric and magnetic fields had been invented/created unabashedly as *intermediaries* between particles which somehow acted on one another at a distance; we preferred to believe that a charged particle created an electric field which then acted on another charged particle rather than the simpler but less mechanical, apparently less real, idea that a charged particle acted across space on another charged particle. Faraday found the visualization of electric and magnetic fields helpful; Maxwell found it possible to put electric and magnetic fields into an elaborate but elegant mathematical theory. With a mathematical structure and empirical verification of what could be measured, electromagnetic believers will swear by the existence of the fields; skeptics can point out correctly that there is no *direct* verification of a field, only its supposed effects on charged particles. Now in QM, what we have called a wave function, a wave, or a particle is some intermediary, a way for us to think of what is clearly a transaction between two parties which we think of a creator/sender and annihilator/receiver. We may be skeptics and say there is *nothing but* the transaction; our attempts at particles, waves, limbo, probabilities are *nothing but* our inventions of the best understood intermediaries we can come up with. But however much we believe is real and what form it takes,

9. Cramer, "Transactional Interpretation."

this transactional interpretation, that ψ is an intermediary traveling forward and backward in time to mediate a transaction does help us make more sense of some things. We may insist that a limbo or superposition state exists, in which *all* the answers to all possible questions cannot exist. We may insist that it has to wait for the question before it has the answer. But signals going backward and forward in time provide a mechanism for not only communicating the question to be asked but having a deal made at the instant of emission, as an observer might call it. If reality is to be accorded this intermediary wave/particle, it *does* have the answer to the question or request which occurs *in the future*. It does not need to have answers to the other questions. And you cannot fool it by a delayed choice of questions; it makes a bargain with the future however far it has to go to make a deal with a future which *occurs*, not a hypothetical future which *might* occur.

Let us freely admit that determinism/indeterminism, randomness/lawfulness, even the real existence of the particle/wave are up for grabs. The time or place or frequency or energy or mass or momentum or spin or whatever you can measure may follow probabilistic rules with apparent random choices in individual cases, but a specific result comes out; no one can say *that* result was impossible, and it was determined somehow. Traditional QM (Copenhagen-style if you wish) had a lottery going on *at the time of* measurement; with this wonderful mechanism of nonlocal interaction, the die can be cast at the beginning and still take the future into account. From its famous historical usage, "the die is cast" can have the double meaning. It can mean indeterminism, as Caesar's *Greek* quoting of his favorite Greek dramatist Menander *Anerriphtho kubos* may be best translated as "*Let* the die be cast." It can mean determinism as the Latin (mis?)translation *Alea iacta est* usually connotes: "the future is determined." The act of crossing the Rubicon, the setting of Caesar's quote, has historically and ironically come to suggest the impossibility of reversing one's action. Here we are positing that Nature provides a reversible path so that the die being cast at the beginning can mean a decision which reflects a future which may include another casting of the die. Apparently random choices at two ends, emission and absorption, can be tied together by two-way signals.

Independently of questions of whether the probability we experience is deeply embedded inextricably in Nature or founded in our ignorance, this nonlocal mechanism of backward/forward waves/particles is satisfying to me for its clarification of the maddening (mis?)use of probability theory in QM. I will be simple about it. Two classical rules of probability are related to our logical categories *or* and *and*: (1) the probability of *one* of two exclusive options A *or* B is the sum of the two probabilities (P(A *or* B) = P(A) + P(B)); (2) the probability of *both* of two independent options A *and* B is the product of the two probabilities (P(A *and* B) = P(A) P(B)). QM seems to mix these in an unfamiliar if not downright wrong manner. The principle of superposition, of limbo in A and/or B, says we *add* ψs, but the probabilities are given by the *square* of ψ, ψ^2, or the product of ψ and ψ. The very concept of superposition confuses our common concept of *or* and *and*, because QM is saying *and* when we are thinking *or* (e.g., the particle is going through this slit *and* that slit when we think the particle is going through this slit *or* that slit). If we think seriously about ψ as a probability, but now of a one-way trip, and a one-way trip as only half of the completed project, then superposition (adding) makes sense of the *ors* and squaring makes sense of the *ands*. It (whatever *it* is) goes all the paths there (all the possible *ors*) *and* all the paths back (all the *ors* again), and all the cross-terms give the characteristic QM interference ideas, which we associate with waves.

$$[P(A) + P(B)]^2 = P^2(A) + P^2(B) + P(A)P(B) + P(B)P(A)$$

Classical physics or classical logic only allows going as A *or* going as B. Quantum physics, if it is a two-fisted fighter capable of, even requiring, a coming as well as a going, has four options: *go A-come A, go B-come B, go A-come B, and go B-come A*. For those of you sophisticated enough to immediately see the problems arising because of complex numbers and negative numbers, let me admit them. As you may be able to see, the problem of complex numbers is a red herring because they always reduce to real numbers when all the terms are in. The negative probabilities sound crazy, but taken as negative *contributions* to a net positive probability, they can bring sense. A catalyst or boost increases probability while a hindrance decreases probability. Feynman has dealt openly with the concept of negative probabilities.[10]

This interpretation of QM does not answer all questions or zip up all the conceptual gaps, but I do find it satisfying. And let me warn you, there are by now *many* interpretations. Wikipedia, let alone the numerous books, will enumerate many and state their supposed stance on several issues such as: realism, completeness, determinism, unique history, collapsing wave functions, existence of universal wave function, and causality. As you can imagine, I have not even mentioned the names of important, brilliant contributors to this discussion. I know and openly admit that this whole matter of QM, theology, and philosophy are three waters over my head (Mark Thrice? or Mark Trine?), but I insist they are not three *different* waters. I admit, even emphasize, that my theology and my QM cannot be separated. My philosophy, in any formal or formulated sense, is so non-existent that it

10 See Feynman, "Negative Probability," in Bohm et al., *Quantum Implications*, Chapter 13, 235–48.

would seem to be irrelevant, but in the depths of unconscious commitments and in the unavoidable bathing, maybe baptism, in cultural catechism, I am both as entrenched and as vulnerable as anyone. I have no serious belief that I can or should convince you of any of my own beliefs. I want you to know, as I have become more aware, that there is an intertwined world of faith, science, philosophy, theology, religion, and everyday mundane experience which beckons us to become entangled and/or disentangled, any and all of which may be to our benefit. And most of all, there is a loving Father at the end of all our trails, saying "Come . . . come home."

Epilogue

> I will give some of my tentative conclusions, but you must make your own.

IF ANY BOOK CAN be considered a call to action or a call to think, then you, the reader, are the epilogist. You will ultimately have the last word, the last act, the final decision. I am under no illusion that the science involved here has any final claim on you or anyone. Not only do I believe that the foundational concepts and basic beliefs undergirding science are uncertain and generally independent of the science itself, I see that any changes due to feedback of the science are slow and also quite uncertain. To tie, as I try, science and theology is to admit that we are in a larger world, a world where everyone's voice and opinion is a legitimate piece of the conversation. We all reflect a set of cultural beliefs and worldviews and we all contribute to the evolving worldview. Whether or not we live in a democratic state which explicitly asks for and uses our input, we will influence the world. And our very basic beliefs are ultimately very important. They may be most obviously fought over by highly influential writers, artists, politicians, etc. but the trickle-down and the trickle-up effects are both important. We can be sold a huge bill of goods one little bit at a time and we can contribute very large social, physical, or spiritual change one little thought and word at a time.

I would love to know and to think that our discussion of these issues has led you toward decisions and contributions to these topics. I am most aware that my decisions and beliefs are still developing and are not likely to be slam-dunked by clear revelatory "answers" from science or from God. I will admit that the *uncertainty* of God's revelation to us in the physical world and the spiritual world is an important belief of mine that you might not share. You may know or expect to know exactly what is right or exactly how it is. When you know, let me know. For what it's worth, let me share where my convictions lie on a number of categories mentioned in the introduction. I must freely admit that I will often not know if my view comes primarily from a theological viewpoint, a scientific viewpoint, or something beyond either, if such is possible.

The duality of continuity/discontinuity seems to be a fundamental part of our existence. Classical physics unconsciously carried the concept of discontinuity in the matter world (atomism or a particle nature of the universe) but became dominated by a continuous world of space, time, causality and deterministic development in space-time. It added the continuous world of fields to the mix. But QM showed us that fields and continuous concepts of space/time motion such as energy, momentum, and angular momentum exhibited the particle-like discontinuous steps-on-a-ladder nature as well. This meant that development of Nature did not only and always go smoothly according to a script; jumps were a necessary part of the program. As far as we can tell, these jumps introduce a necessary element of chance, choice, randomness, creation, and leaps of faith. Some may wish to believe that God and Nature follow exact differential equations in a machine-like perfect manner; I prefer to believe that there is choice, chance, grace, personal decisions, and new creations from top to bottom of the natural/supernatural order.

Some of you will reject chance and randomness out of hand; it is impossible with your view of God or Nature. Nature and God are perfectly determined in every detail. That's good for a machine, I guess, but, for me, I choose to believe in a personal God who built chance in, at least as early as he made mankind in his image. When he gave commands, it was implicit that those commands could be disobeyed. When it involves persons, I think my decision is due to my theological belief that God relates to something made in his image, something, someone, who has choices and is condemned (according to Sartre's thinking) to make them. When it comes to inanimate matter, I suppose I could believe in a physical world which has no choice, a world which does not participate in the purposeful plane of personhood, acting only as a substrate for life. I don't like the implied segregated duality, but it seems possible. It seems to me that QM indicates a real realm of randomness. Randomness is so hard to define that we may never know what it means or how we can test it. But QM seems to indicate randomness and QM phenomena pass our best tests for randomness. We can produce pseudo-random numbers in many ways; QM seems to be the best random generator we have. Gambling businesses set up games which depend on randomness; if the machines (cards, lotteries, dice, roulette wheels, etc.) work correctly, i.e., randomly, the house makes money. One definition of randomness is that you, the gambler, cannot beat the house. You cannot find any non-random feature to exploit. Those who think God allows/uses randomness generally think he is an honest gambler; the dice-throwing God does not play with loaded dice.[1]

We, in theology *or* science, have the task of trying to see how God/Nature can be meaningful with a substrate of randomness. Generally the scale of large (compared to atomic sizes) objects with its averages over many atomic events take quite good care of the problem. After all, we approached this whole problem from a macroscopic world that everyone in the late nineteenth century believed was determined. Just as thermodynamics was able to have strong laws while hiding its eyes from detail which it could not see anyway, the statistics of large numbers produce a very good approximation to determinism. Thus a statistical causality, the very accurate shaping of probabilistic distributions, works just about as well as the iron-clad laws supposed in classical physics. FAPP, "for all practical purposes," Nature *looks* determined.

But Schrödinger's cat dies *randomly* from *one* quantum event and a bomb that could annihilate the whole Earth could just as easily be triggered by *one* event, even one bit of information. If God allows chance and choice anywhere, how can he control these individual events? This is the question that is pertinent for theologians and physicists alike. How could random events produce any lawful relationships? That was a major background question of EPR. Two *independent random* events in the two arms of an EPR experiment cannot produce correlated results. But QM says the events are correlated and it proves true in the lab. 314,159,265,358,97 . . . (out to millions of places) and 203,048,154,247,86 . . . (out to millions of places) can both pass all the tests for randomness of ten digits, but the quick among you may note that subtracting the second from the first, integer by integer (add 10 if you get a negative), gives a quite un-random number 111,111,111,111,11 . . . (out to millions of places).[2] The quick and unbelieving, let's call them a-randomists, will also gleefully point out that the first (and therefore the second) number is not random at all but is essentially the number π, perfectly determined by Nature and by mathematical algorithms to calculate it. It appears we may always have this problem of not knowing if something is random, but we know that the random-appearing answers are correlated where QM produces entanglements, "twins" which show that the two (or more) supposedly random sequences are not independent. Quantum nonlocality allows a *meaningful* non-random signal in the cross-correlation of apparently, maybe really, random signals.

1. Pagels, *Cosmic Code*, 91.

2. Ibid., 88–89.

There appears to be a possible nonlocal, faster-than-the-speed-of-light communication channel, which will excite some and raise important objections for others. Although we can see the quite non-random correlations, quite capable of carrying information, after the fact, the randomness of each of the signals prevents reading any message from either signal alone. It is as though you may receive the message nonlocally, faster than light can get there, but it is coded and the decoding signal is required and will not get there faster than light. It appears that both randomness and nonlocality figured strongly in Einstein's objections to QM; it also appears that randomness prevents the nonlocality which appears strongly at the quantum level from becoming nonlocality that we can experience and exploit. The paradoxes arising from faster-than-light and backward-in-time signals do not seem to arise on our level. Killing your own grandparent and thus preventing your own birth does not seem to be one of the choices that are possible for Sartre or any of us. Heinz Pagels would say that randomness is a trick that "the God that (sic) plays dice" uses to prevent "*real* nonlocality" (my emphasis, his word).[3] For years, John G. Cramer thought he had a way (with the retrocausality implicit in delayed-choice experiments) of sending a signal backward in time, even if only a few microseconds. You might want to look at his experience for a much more complete, much more sophisticated explanation than we can do here.[4]

The nonlocality produces curious and fascinating possibilities in questions of the nature of time and retrocausality (or teleology), but the clear non-randomness of cross-correlations of seemingly random signals helps us understand how lawful, meaningful relationships can be the result of randomness. Randomness may be the fabric of a curtain by which God limits our knowledge, whether that curtain is transparent to him or not. But two random fabrics woven together can produce a clearly non-random picture of cross-correlations. Because our minds seek patterns and can find them in very subtle situations, we can make random patterns in which the combining of the two views from our two eyes will be seen as a clearly non-random pattern (just as stereoscopic pictures seen separately by the two eyes because of literal physical separation or by polarizing glasses can produce the third dimension in 3-D pictures and movies). This is illustrated by the random pattern shown on the next page. These can be generated on-line and used for free.[5] Adjusting one's eyes to see these patterns can be difficult for some of us. The usual instruction is to put your face right up to the paper

3. Ibid., 151.

4. https://faculty.washington.edu/jcramer/

5. http://www.easystereogrambuilder.com/

and think of looking through the picture to a point behind, then move your face back slowly without changing the focus. You should then feel like you are seeing something behind the picture, a 3-D effect. It might help to concentrate on either the top or the bottom of the picture until the image pops out for you. It also may help if you know the image shows some letters of the alphabet. When you see them, you will know what they represent.

We have clearly suggested that we, the questioners, can have an effect on Nature by the questions we ask, even when we delay our choice of questions until we think Nature must have made her choice. Again the randomness protects Nature from our imposition of *specific* results; the natural laws are manifest in correlations. For example, our choice of settings in Mermin's impossible machine can never determine the color (red or green) of the flashing light; it only affects the *relationship* between the two colors flashed in the two arms. Quantum physics does allow us (or God) to get any answer we wish if we can know the result of a previous (or other arm) measurement (which does not destroy the photon). If a photon says "yes" to our polarization question, we can ask the same question if we want a "yes" or the complementary question (90^0 with respect to the first) if we want a "no." This may really sound cheap but a God who knows the past without a destructive experiment can get any message (meaningful sequence of answers) he wants by his choice of questions. Those of you concerned about protecting your God from any slightest hint of chance or lack of knowledge of the future will clearly not be satisfied by that.

The ability of God/Nature/experimenters to control the results by choices of measurements is much more sophisticated than the cheap solution of asking the questions to which we already know the answers. Let's give an example in our simplest two-state system of polarizations. If a photon exists, we say that it *does not have* a polarization state because (1) we do not know what angle a polarizing filter or beam-splitter will be set at and (2) the answer to the two-choice question will be random. This is the possible situation of ignorance where God may allow the free will of a questioner and the free will (randomness?) of the photon. Suppose God has a specific question (angle of polarization) and answer which he wants or is asked to get for that photon which already *has* a different polarization because of a previous measurement. By making a continuous set of measurements starting at the now-known random beginning and gradually moving to the desired final answer, God (or any other experimenter capable of such measurements) can get the final answer. This is a well-known classical result for *waves* which carries over into quantum phenomena for a *single photon*.

Recall that a polarized photon will have $\cos^2\theta$ chance of going through a θ-polarizing filter (i.e., of *being* a θ-photon). In the extreme case, an x-axis photon has *no* chance of being a y-axis photon, going through a polarizer at 90^0. But if you place a 45^0 filter between, half will go through it and half of those will go through the 90^0 filter. The intermediate measurement which *cuts* half of the photons *increases* (from 0 to (1/4)) the output, the chances of going through the 90^0 filter. Two intermediate filters (at 30^0 and 60^0) increase it further to (27/64). More and more intermediate filters spaced equally (in angles) increase the chances until finally a continuous set of measurements allows 100 percent to go through. Even though such measurements give *random* results for an individual case, the shape ($\cos^2\theta$) of the distribution allows the choice of measurements to take absolute control of the final result.

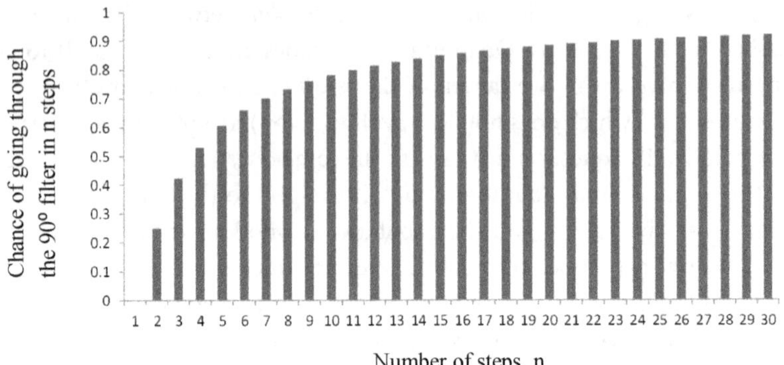

This same kind of phenomenon can be reproduced in time dependence for energy transitions in atoms (absorption or emission of particles). In a spontaneous emission (like radioactivity), there is a characteristic time, the half-life, at which the probability for the emission to have occurred is (1/2). If we have a way of checking an individual atom to ask if the transition has occurred, the measurement, the collapse of the wave function in QM terminology, can reset the clock. By checking more times, more often, you increase the half-life. A continuously asked atom will never make the transition. This is a serious and real quantum effect, a real manifestation of our joking aphorism that a watched pot never boils. Because this has all the makings of the ancient Zeno paradoxes which claimed motion could not exist, this is sometimes called the quantum *Zeno* effect.[6]

Randomness would seem to take all the steam out of the observer's supposed effect on the measurement, but the statistical causality with QM distributions allows perfect control by the continuous choices of a watchful supervisor. A Shepherd who keeps his watchful eye on his flock can allow randomness and free will and still not lose a single sheep. The Final Judgment is the big crucial measurement, but God did not leave us *only* with a mechanistic deterministic causality or the opportunity to stray. He keeps close tabs on the free creatures among apparent machines and apparently free (even random) Nature and generally draws us to our *telos*, his bosom at the great feast of the Lamb. Nature may seem like an inexorable machine, a grand roulette wheel, or a capricious God with a portfolio of magic tricks, but if we look with eyes of faith, I believe we will still see a caring Father awaiting us with open arms.

6. Greenstein and Zajonc, *Quantum Challenge*, 231–37.

A God who breaks into the physical realm with a smashing incomprehensible God/man reality can break the object/subject duality, can provide freedom and yet keep a lawful world, can supplement our woefully weak concept of reality in being and causality with a huge dose of becoming and teleology, can create meaning out of apparent randomness, can create life out of erstwhile machines, and can maintain absolutes and metanarratives when our imagined narratives and cherished absolutes collapse around us. The unity and wholeness of the universe and the unity and wholeness of our scientific and religious experience of the universe do not require our arguments or intellectual support, but they must be believed and entered into.

Appendix 1: Quantization condition

IF AN ELECTRON GOES in a circular orbit around a heavy nucleus, the force law (Coulomb's law of electric forces) produces a velocity proportional to $1/\sqrt{r}$. A constant angular momentum produces a velocity proportional to $1/r$. In the fashion of good theoretical physics, using natural units will set the proportionality constants to 1. Thus $v_F = 1/\sqrt{r}$ (velocity determined by the force law) and $v_L = 1/r$ (velocity determined by L, the angular momentum).

There is of course only one velocity. Thus $1/\sqrt{r} = 1/r$, which even you and I can solve, since the answer is clearly $r = 1$. This is a real natural unit, called the Bohr radius, we'll label it r_1, and of course comes out to be basically the size of an atom. Thus angular momentum chosen at some natural value (clearly related to h) produces the magical orbit size which is stable, not only mechanically, but stable against the catastrophic electromagnetic radiation loss and the consequent collapse of the atom. If, in these natural units, $v_L = n/r$ where $n = 1, 2, 3, \ldots$, one easily obtains a series of radii which act as *stationary states.*

$$n/r_n = 1/\sqrt{r_n}\,.$$

$$n^2/r_n^2 = 1/r_n.$$

Therefore, $r_n = n^2$ in natural units, or $r_n = n^2 r_1$.

In our solar orbits, the continuity condition (our assumption) says that any and all radii are possible. Here in our first inklings of quantum mechanics, discrete non-continuous values being the only possible values for radii, angular momenta, velocities, and energies, we see that the hints in Planck's work of the *particle-ization* of electromagnetic waves are carried over into many other continuous variables. The picture below shows the graphing of the velocity produced by the force law, $v_F = 1/\sqrt{r}$, and the first few values of velocity forced by angular momentum *quantization*: $v_1 = 1/r$, $v_2 = 2/r$, and $v_3 = 3/r$. *The crossing points are* $1/r_1 = 1$, $1/r_2 = 1/4$, $1/r_3 = 1/9$, etc.

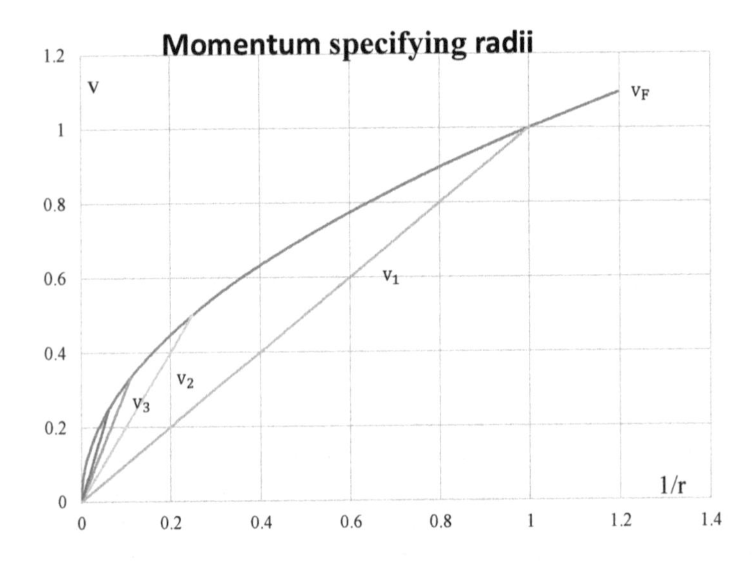

This is easy enough because we have faked the numbers (by choosing natural units) and have known the pattern in advance. If you want to try the next harder step, find the pattern in the following wavelengths of the hydrogen spectrum: 656.3 nm, 486.1 nm, 434.1 nm, 410.2 nm, 397.0 nm, 388.9 nm, 383.5 nm, called the Balmer series. (I went to the trouble of doing it myself.) One hint should be sufficient if you will work at it. Take the reciprocal of these numbers (converting wavelengths to frequencies) and look for relationships to the series above remembering that energy is related to v^2 (kinetic energy = $(1/2)mv^2$).

The next harder step was to find this small set of wavelengths out of the thousands of wavelengths of spectrographic data. After the fact, they can look reasonably easy. Approaching it as it occurred historically, it seems miraculous, magical, or mysterious. This required some genius, perseverance, timing, luck, etc., and this was most definitely not Einstein's type of genius.

Appendix 2

SUPPOSE THERE IS AN original message of 0s and 1s, and transcriptions (statistically) have an error rate of 0.2. Let us do two independent transcriptions, shown below. Original message is bold, shown in middle; transcriptions are shown above and below.

0 0 1 0 0 1 1 1 0 1 0 1 1 1 0 1 0 0 0 1

0 0 1 1 1 1 1 1 0 1 1 1 1 1 0 1 0 0 1 1

0 0 1 1 1 1 1 1 0 1 1 1 1 1 0 1 0 0 0 0

There are 4 errors in the upper transcription and 2 errors in the lower transcription. Now suppose that the original is lost or was never known. Comparing the two *transcribed* messages (below), we can see that between the two, there appear only 4 transcription errors. The double error (at position 19) does not show as an error. Thus E(2) will always be less than E(1) + E(1) = 2E(1).

0 0 1 0 0 1 1 1 0 1 0 1 1 1 0 1 0 0 0 1

0 0 1 1 1 1 1 1 0 1 1 1 1 1 0 1 0 0 0 0

QM breaks this rule and thus *must not have* the "original" message and the so-called independent transcriptions are not independent. As stated in the text, E(2) is about equal to 4 E(1) for quantum-mechanical measurements with low error rates. Imagine two monks producing two copies of a manuscript and producing twice as many errors as they would when the manuscripts were done independently. You would be convinced that something fishy was going on. In QM, we are convinced that the "original" message does not exist; there is some limbo state and isolation of the two supposedly independent measurements is impossible. Imagine locking the monks in isolation quarters and finding they can produce manuscripts that prove they are still capable of this kind of fishy collaboration. This is what QM finds. That is, this is how the quantum world behaves.

Bibliography

AIP (American Institute of Physics), Center for History of Physics. Available on-line.

Albert, David Z. *Quantum Mechanics and Experience.* Cambridge: Harvard University Press, 1992.

Alexander, Amir R. *Infinitesimal: How a Dangerous Mathematical Theory Shaped the Modern World.* London: Oneworld, 2015.

Allison, Gregg R., and Wayne A. Grudem. *Historical Theology: An Introduction to Christian Doctrine: A Companion to Wayne Grudem's Systematic Theology.* Grand Rapids: Zondervan, 2011.

Atkins, P. W. *Four Laws that Drive the Universe.* Oxford: Oxford University Press, 2007.

Atmanspacher, Harald, and Hans Primas. *Recasting Reality: Wolfgang Pauli's Philosophical Ideas and Contemporary Science.* Berlin: Springer, 2009.

Auden, W. H. *A Certain World: A Commonplace Book.* New York: Viking, 1970.

Bacon, Francis, Peter Urbach, John Gibson, and Francis Bacon. *Novum Organum; with Other Parts of the Great Instauration.* Chicago: Open Court, 1994.

Baggott, J. E. *Beyond Measure: Modern Physics, Philosophy, and the Meaning of Quantum Theory.* Oxford: Oxford University Press, 2004.

———. *The Meaning of Quantum Theory: A Guide for Students of Chemistry and Physics.* Oxford: Oxford University Press, 1992.

Baierlein, Ralph. *Atoms and Information Theory: An Introduction to Statistical Mechanics.* San Francisco: Freeman, 1971.

Barbour, Ian G. *Issues in Science and Religion.* Englewood Cliffs, NJ: Prentice-Hall, 1966.

———. *Religion in an Age of Science.* San Francisco: Harper & Row, 1990.

Barfield, Owen. *Saving the Appearances: A Study in Idolatry.* New York: Harcourt, Brace & World, 1965.

Barr, Stephen M. *Modern Physics and Ancient Faith.* Notre Dame, IN: University of Notre Dame Press, 2003.

Bartholomew, David J. *God, Chance, and Purpose: Can God Have It Both Ways?* Cambridge: Cambridge University Press, 2008.

Bell, J. S. *Speakable and Unspeakable in Quantum Mechanics: Collected Papers on Quantum Philosophy.* Cambridge: Cambridge University Press, 1987.

Beller, Mara. *Quantum Dialogue: The Making of a Revolution.* Chicago: University of Chicago Press, 1999.

Bernstein, Jeremy. *Quantum Leaps.* Cambridge: Belknap Press of Harvard University Press, 2009.

———. *Quantum Profiles*. Princeton, NJ: Princeton University Press, 1991.

Bohm, David. *Causality and Chance in Modern Physics*. Philadelphia: University of Pennslyvania Press, 1999.

———. *Quantum Theory*. New York: Prentice-Hall, 1951.

———. "A Suggested Interpretation of the Quantum Theory in Terms of 'Hidden Variables,' I and II." *Physical Review* 85 (1952) 165–79 and 180–93.

Bohm, David, Charles Joseph Biederman, and P. Pylkkänen. *Bohm-Biederman Correspondence, Volume 1*. 1999. Reprint. London: Routledge, 2009.

Bohm, David, and Lee Nichol. *The Essential David Bohm*. London: Routledge, 2003.

Bohm, David, B. J. Hiley, and F. David Peat. *Quantum Implications: Essays in Honour of David Bohm*. New York: Routledge & Kegan Paul in association with Methuen, 1987.

Bohr, Niels. *Atomic Physics and Human Knowledge*. New York: Wiley. 1958.

———. *The Philosophical Writings of Niels Bohr*. Woodbridge, CT: Ox Bow, 1987.

Bohr, Niels, L. Rosenfeld, Erik Rüdinger, and Finn Aaserud. *Collected Works*. Amsterdam: North-Holland, 1972.

Bolles, Edmund Blair. *Einstein Defiant: Genius versus Genius in the Quantum Revolution*. Washington, DC: Joseph Henry, 2004.

Born, Max. *My Life & My Views*. New York: Scribner, 1968.

———. *Natural Philosophy of Cause and Chance: Being the Waynflete Lectures, Delivered in the College of St. Mary Magdalen, Oxford, in Hilary Term, 1948, Together with a New Essay, "Symbol and Reality."* New York: Dover, 1964.

Born, Max. *Physics in my Generation*. New York: Springer-Verlag, 1969.

Born, Max, and Emil Wolf. *Principles of Optics*. London: Pergamon, 1959.

Boyd, Gregory A. *God of the Possible: A Biblical Introduction to the Open View of God*. Grand Rapids: Baker, 2000.

Brennan, Richard P. *Heisenberg Probably Slept Here: The Lives, Times, and Ideas of the Great Physicists of the Twentieth Century*. New York: Wiley, 1997.

Broglie, Louis de. *Physics and Microphysics*. New York: Pantheon, 1955.

Broglie, Louis de, William Charles Price, Seymour S. Chissick, and Tom Ravensdale. *Wave Mechanics: The First Fifty Years. A Tribute to Professor Louis de Broglie, Nobel Laureate, on the Fiftieth Anniversary of the Discovery of the Wave Nature of the Electron*. New York: Wiley, 1973.

Brooke, John Hedley. *Science and Religion: Some Historical Perspectives*. Cambridge: Cambridge University Press, 1991.

Bruce, Colin. *Schrödinger's Rabbits: The Many Worlds of Quantum*. Washington, DC: Joseph Henry, 2004.

Burchfield, Joe D. *Lord Kelvin and the Age of the Earth*. Chicago: University of Chicago Press, 1990.

Cahill, Thomas. *The Gifts of the Jews: How a Tribe of Desert Nomads Changed the Way Everyone Thinks and Feels*. New York: Talese, 1998.

Campbell, Lewis, and William Garnett. *The Life of James Clerk Maxwell*. New York: Johnson Reprint Corp., 1969.

Capra, Fritjof. *The Tao of Physics: An Exploration of the Parallels between Modern Physics and Eastern Mysticism*. Berkeley: Shambhala, 1975.

———. *The Turning Point: Science, Society, and the Rising Culture*. New York: Simon and Schuster, 1982.

Caspar, Max. *Kepler*. London: Abelard-Schuman, 1959.

Cassidy, David C. *Uncertainty: The Life and Science of Werner Heisenberg*. New York: Freeman, 1992.

Chester, Marvin. *Primer of Quantum Mechanics*. New York: Wiley, 1987.

Colodny, Robert Garland, and Norwood Russell Hanson. *Beyond the Edge of Certainty: Essays in Contemporary Science and Philosophy*. Englewood Cliffs, NJ: Prentice-Hall, 1965.

Compton, Arthur Holly. *The Cosmos of Arthur Holly Compton*. New York: Knopf, 1968.

Cramer, John G. "The Transactional Interpretation of Quantum Mechanics." *Reviews of Modern Physics* 58 (1986) 647–88.

———. Website. https://faculty.washington.edu/jcramer/.

Crease, Robert P., and Charles C. Mann. *The Second Creation: Makers of the Revolution in Twentieth-Century Physics*. New York: MacMillan, 1986.

Crowther, J. G. *Men of Science: Humphry Davy, Michael Faraday, James Prescott Joule, William Thomson, James Clerk Maxwell*. 1936. Reprint. New York: Norton, 1973.

Cushing, James T. *Philosophical Concepts in Physics: The Historical Relation between Philosophy and Scientific Theories*. Cambridge: Cambridge University Press, 1998.

———. *Quantum Mechanics: Historical Contingency and the Copenhagen Hegemony*. Chicago: University of Chicago Press, 1994.

Dirac, P. A. M. *The Principles of Quantum Mechanics*. Oxford: Clarendon, 1983.

Eddington, Arthur Stanley. *The Nature of the Physical World*. New York: Macmillan, 1928.

Ehrenfest, P., and M. J. Klein. *Collected Scientific Papers*. Amsterdam: North Holland, 1959.

Einstein, Albert, Anna Beck, and Peter Havas. *The Collected Papers of Albert Einstein: English Translation*. Princeton, NJ: Princeton University Press, 1987.

Einstein, Albert, Max Born, and Hedwig Born. *The Born-Einstein Letters: Friendship, Politics, and Physics in Uncertain Times: Correspondence between Albert Einstein and Max and Hedwig Born from 1916 to 1955 with Commentaries by Max Born*. Basingstoke, UK: Macmillan, 2005.

Einstein, Albert, B. Podolsky, and N. Rosen. "Can Quantum Mechanical Description of Physical Reality Be Considered Complete?" *Physical Review* 47 (1935) 777–80.

Einstein, Albert, and John J. Stachel. *The Collected Papers of Albert Einstein*. Princeton, NJ: Princeton University Press, 1987.

Enz, Charles P. *No Time to be Brief: A Scientific Biography of Wolfgang Pauli*. Oxford: Oxford University Press, 2002.

Espagnat, Bernard d'. *In Search of Reality*. New York: Springer-Verlag, 1983.

———. *On Physics and Philosophy*. Princeton: Princeton University Press, 2006.

———. *Veiled Reality: An Analysis of Present-Day Quantum Mechanical Concepts*. Reading, MA: Addison-Wesley, Advanced Book Program, 1995.

Evans, C. Stephen. *Kierkegaard on Faith and the Self-collected Essays*. Waco, TX: Baylor University Press, 2006.

Faries, Dillard W. "A Personal God, Chance, and Randomness in Quantum Physics." *Perspectives on Science and Christian Faith* 66.1 (2014) 13–22.

Farmelo, Graham. *The Strangest Man: the Hidden Life of Paul Dirac, Mystic of the Atom*. New York: Basic, 2009.

Faye, Jan. *Niels Bohr: His Heritage and Legacy: An Anti-realist View of Quantum Mechanics*. Dordrecht: Kluwer Academic, 1991.

Ferguson, Kitty. *The Music of Pythagoras: How an Ancient Brotherhood Cracked the Code of the Universe and Lit the Path from Antiquity to Outer Space.* New York: Walker, 2008.

———. *Tycho & Kepler: The Unlikely Partnership That Forever Changed Our Understanding of the Heavens.* New York: Walker, 2002.

Feynman, Richard P., and Albert R. Hibbs. *Quantum Mechanics and Path Integrals.* New York: McGraw-Hill, 1965.

Feynman, Richard P., Robert B. Leighton, and Matthew L. Sands. *The Feynman Lectures on Physics.* Reading, MA: Addison-Wesley, 1963.

Feynman, Richard P., Robert B. Leighton, and Matthew L. Sands. *Six Easy Pieces: Essentials of Physics, Explained by Its Most Brilliant Teacher.* Reading, MA: Addison-Wesley, 1995.

Fine, Arthur. *The Shaky Game: Einstein, Realism, and the Quantum Theory.* Chicago: University of Chicago Press, 1986.

Flood, Raymond, Mark McCartney, and Andrew Whitaker. *Kelvin: Life, Labours and Legacy.* Oxford: Oxford University Press, 2008.

Folse, Henry J. *The Philosophy of Niels Bohr: The Framework of Complementarity.* Amsterdam: North-Holland, 1985.

Forbes, Nancy, and Basil Mahon. *Faraday, Maxwell, and the Electromagnetic Field: How Two Men Revolutionized Physics.* Albany, NY: Prometheus, 2014.

Frayn, Michael. *Copenhagen.* New York: Anchor, 2000.

Gilder, Louisa. *The Age of Entanglement: When Quantum Physics Was Reborn.* New York: Knopf, 2008.

Goldman, Robert N., and Albert Einstein. *Einstein's God: Albert Einstein's Quest as a Scientist and as a Jew to Replace a Forsaken God.* Northvale, NJ: Aronson, 1997.

Greenspan, Nancy Thorndike. *The End of the Certain World: The Life and Science of Max Born, the Nobel Physicist Who Ignited the Quantum Revolution.* New York: Basic, 2005.

Greenstein, George, and Arthur Zajonc. *The Quantum Challenge: Modern Research on the Foundations of Quantum Mechanics.* Sudbury, MA: Jones and Bartlett, 1997.

Griffiths, R. B. *Consistent Quantum Theory.* Cambridge: Cambridge University Press, 2002.

Hamlyn, D. W. *A History of Western Philosophy.* Harmondsworth, UK: Viking, 1987.

Hecht, Eugene. *Optics, Second Edition.* Reading, MA: Addison-Wesley, 2002.

Heilbron, J. L. *The Dilemmas of an Upright Man: Max Planck as Spokesman for German Science.* Berkeley: University of California Press, 1986.

Heisenberg, Elisabeth. *Inner Exile: Recollections of a Life with Werner Heisenberg.* Boston: Birkhäuser, 1984.

Heisenberg, Werner. *The Physical Principles of the Quantum Theory.* New York: Dover, 1950.

———. *Physics and Beyond: Encounters and Conversations.* New York: Harper & Row, 1972.

———. *Physics and Philosophy: The Revolution in Modern Science.* New York: Harper, 1958.

Hendry, John. *The Creation of Quantum Mechanics and the Bohr-Pauli Dialogue.* Dordrecht: Reidel, 1984.

Herman, Arthur. *The Cave and the Light: Plato versus Aristotle and the Struggle for the Soul of Western Civilization.* New York: Random House, 2013.

Hodgson, P. E. *Theology and Modern Physics*. Aldershot, UK: Ashgate, 2005.
Hoffman, Banesh. *The Strange Story of the Quantum*. London: Penguin, 1963.
Hoffman, Banesh, with Helen Dukas. *Albert Einstein, Creator and Rebel*. New York: Viking, 1973.
Isaacson, Walter. *Einstein: His Life and Universe*. New York: Simon and Schuster, 2007.
Jaki, Stanley L. *The Absolute beneath the Relative and Other Essays*. Lanham, MD: University Press of America, 1988.
———. *Chance or Reality and Other Essays*. Lanham, MD: University Press of America, 1986.
———. *God and the Cosmologists*. Washington, DC: Regnery Gateway, 1989.
———. *The Relevance of Physics*. Chicago: University of Chicago Press, 1966.
———. *The Road of Science and the Ways to God*. Chicago: University of Chicago Press, 1978.
Jammer, Max. *The Conceptual Development of Quantum Mechanics*. New York: McGraw-Hill, 1966.
———. *Einstein and Religion: Physics and Theology*. Princeton, NJ: Princeton University Press, 1999.
———. *The Philosophy of Quantum Mechanics: The Interpretations of Quantum Mechanics in Historical Perspective*. New York: Wiley, 1974.
Jeans, James. *Physics & Philosophy*. Cambridge: Cambridge University Press, 1943.
Jones, Roger S. *Physics as Metaphor*. Minneapolis: University of Minnesota Press, 1982.
Jones, Sheilla. *The Quantum Ten: A Story of Passion, Tragedy, Ambition and Science*. Oxford: Oxford University Press, 2008.
Jung, C. G., and Wolfgang Pauli. *Naturerklärung und Psyche. Synchronizität als ein Prinzip akausaler Zusammenhänge*. Zürich: Rascher, 1952.
Jung, C. G., and Wolfgang Pauli. *The Interpretation of Nature and the Psyche*. London: Routledge & Kegan Paul, 1955.
Kaku, Michio. *Physics of the Impossible: A Scientific Exploration into the World of Phasers, Force Fields, Teleportation, and Time Travel*. New York: Doubleday, 2008.
Kanigel, Robert. *The Man Who Knew Infinity: A Life of the Genius, Ramanujan*. New York: Scribner's Sons, 1991.
Kargon, Robert H. *The Rise of Robert Millikan: Portrait of a Life in American Science*. Ithaca, NY: Cornell University Press, 1982.
Kepler, Johannes. *The Six-Cornered Snowflake*. Oxford: Clarendon, 1966.
Kepler, Johannes, and Carola Baumgardt. *Johannes Kepler: Life and Letters*. New York: Philosophical Library, 1951.
Klein, Martin J. *Paul Ehrenfest*. Amsterdam: North-Holland, 1970.
Kocher, Carl Alvin. "Polarization Correlation of Photons Emitted in an Atomic Cascade." PhD diss., University of California, Berkeley, 1967.
Koestler, Arthur. *The Watershed: A Biography of Johannes Kepler*. Garden City, NY: Anchor, 1960.
Kragh, Helge. *Quantum Generations: A History of Physics in the Twentieth Century*. Princeton, NJ: Princeton University Press, 1999.
Kuhn, Thomas S. *Black-Body Theory and the Quantum Discontinuity, 1894–1912*. Oxford: Clarendon, 1978.
Landé, Alfred. *From Dualism to Unity in Quantum Physics*. Cambridge: Cambridge University Press, 1960.

Landé, Alfred, Wolfgang Yourgrau, and Alwyn Van der Merwe. *Perspectives in Quantum Theory: Essays in Honor of Alfred Landé*. Cambridge: MIT Press, 1971.

Laurikainen, Kalervo Vihtori. *Beyond the Atom: The Philosophical Thought of Wolfgang Pauli*. Berlin: Springer-Verlag, 1988.

———. *The Message of the Atoms: Essays on Wolfgang Pauli and the Unspeakable*. Berlin: Springer, 1997.

L'Engle, Madeleine, and Luci Shaw. *Wintersong: Christmas Readings*. Wheaton, IL: Shaw, 1996.

Lewis, C. S., and Paul F. Ford. *Yours, Jack: Spiritual Direction from C. S. Lewis*. New York: Harper One, 2008.

Lindley, David. *Boltzmann's Atom: The Great Debate that Launched a Revolution in Physics*. New York: Free, 2001.

———. *The End of Physics: The Myth of a Unified Theory*. New York: Basic, 1993.

———. *Uncertainty: Einstein, Heisenberg, Bohr, and the Struggle for the Soul of Science*. New York: Doubleday, 2007.

Lindorff, David P. *Pauli and Jung: The Meeting of Two Great Minds*. Wheaton, IL: Quest, 2004.

Mach, Ernst. *The Science of Mechanics: A Critical and Historical Account of Its Development*. LaSalle, IL: Open Court, 1960.

Madison, Gary Brent. *The Hermeneutics of Postmodernity: Figures and Themes*. Bloomington, IN: Indiana University Press, 1988.

Margenau, Henry. *The Nature of Physical Reality: A Philosophy of Modern Physics*. New York: McGraw-Hill, 1950.

———. *Open Vistas: Philosophical Perspectives of Modern Science*. New Haven: Yale University Press, 1961.

McFadden, Johnjoe. *Quantum Evolution*. New York: Norton, 2001.

McKay, Donald M. *Christianity in a Mechanistic Universe and Other Essays*. Downer's Grove, IL: IVP, 1966.

Mehra, Jagdish, and Helmut Rechenberg. *The Historical Development of Quantum Theory*. New York: Springer-Verlag, 1982.

Mermin, N. David. "Is the Moon There When Nobody Looks? Reality and the Quantum Theory." *Physics Today* 38.4 (1985) 38–47.

Mooney, Christopher F. *Theology and Scientific Knowledge: Changing Models of God's Presence in the World*. Notre Dame, IN: University of Notre Dame Press, 1996.

Moore, Walter John. *Schrödinger, Life and Thought*. Cambridge: Cambridge University Press, 1989.

Morris, Tim, and Donald N. Petcher. *Science & Grace: God's Reign in the Natural Sciences*. Wheaton, IL: Crossway, 2006.

Mumford, Lewis. *My Works and Days: A Personal Chronicle*. New York: Harcourt Brace Jovanovich, 1979.

Newton, Isaac, I. Bernard Cohen, and Anne Miller Whitman. *The Principia: Mathematical Principles of Natural Philosophy*. Berkeley: University of California Press, 1999.

Nobelstiftelsen (Nobel Foundation). *Nobel Lectures: Physics, 1901–1921*. Amsterdam: Elsevier, 1967.

Nobelstiftelsen (Nobel Foundation). *Nobel Lectures: Physics, 1922–1941*. Amsterdam: Elsevier, 1965.

Nobelstiftelsen (Nobel Foundation). *Physics, 1942–1962*. Singapore: World Scientific, 1998.

Oppenheimer, J. Robert. *Atom and Void: Essays on Science and Community*. Princeton, NJ: Princeton University Press, 1989.

Pagels, Heinz R. *The Cosmic Code: Quantum Physics as the Language of Nature*. New York: Simon and Schuster, 1982.

Paine, Albert Bigelow. *Mark Twain: A Biography: The Personal and Literary Life of Samuel Langhorne Clemens*. New York: Harper, 1912.

Pais, Abraham. *"Subtle is the Lord—": The Science and the Life of Albert Einstein*. Oxford: Oxford University Press, 1982.

Parker, Barry R. *Einstein: The Passions of a Scientist*. Amherst, NY: Prometheus, 2003.

Pauli, Wolfgang. *Collected Scientific Papers*. New York: Interscience, 1964.

Pauli, Wolfgang, C. A. Meier, Charles P. Enz, Markus Fierz, C. G. Jung, and David Roscoe. *Atom and Archetype: The Pauli/Jung Letters, 1932–1958*. Princeton, NJ: Princeton University Press, 2001.

Peat, F. David. *From Certainty to Uncertainty: The Story of Science and Ideas in the Twentieth Century*. Washington, DC: Joseph Henry, 2002.

Petruccioli, Sandro, and Ian McGilvray. *Atoms, Metaphors, and Paradoxes: Niels Bohr and the Construction of a New Physics*. Cambridge: Cambridge University Press, 1994.

Pirsig, Robert M. *Zen and the Art of Motorcycle Maintenance: An Inquiry into Values*. New York: Morrow, 1974.

Planck, Max. *Scientific Autobiography and Other Papers*. New York: Philosophical Library, 1949.

Plotnitsky, Arkady. *Complementarity: Anti-Epistemology after Bohr and Derrida*. Durham, NC: Duke University Press, 1994.

———. "From Como to Copenhagen: The Beginnings and Ends of Complementarity." *AIP Conference Proceedings* 962.1 (2007) 185–94.

———. *The Principles of Quantum Theory, from Planck's Quanta to the Higgs Boson: The Nature of Quantum Reality and the Spirit of Copenhagen*. Cham, Switzerland: Springer, 2016.

———. *Reading Bohr: Physics and Philosophy*. Dordrecht, the Netherlands: Springer, 2006.

Polanyi, Michael. *Personal Knowledge: Towards a Post-Critical Philosophy*. Chicago: University of Chicago Press, 1958.

Polkinghorne, John C. *Quantum Physics and Theology: An Unexpected Kinship*. New Haven: Yale University Press, 2007.

———. *The Quantum World*. London: Longman, 1984.

Pollard, William G. *Chance and Providence: God's Action in a World Governed by Scientific Law*. New York: Scribner's Sons, 1958.

Powers, Thomas. *Heisenberg's War: The Secret History of the German Bomb*. New York: Knopf, 1993.

Price, Huw. *Time's Arrow & Archimedes' Point: New Directions for the Physics of Time*. New York: Oxford University Press, 1996.

Przibram, Karl, Erwin Schrödinger, Max Planck, Albert Einstein, and H. A. Lorentz. *Letters on Wave Mechanics: Schrödinger, Planck, Einstein, Lorentz*. New York: Philosophical Library, 1967.

Rosenblum, Bruce, and Fred Kuttner. *Quantum Enigma: Physics Encounters Consciousness*. New York: Oxford University Press, 2011.

Rozental, S. *Niels Bohr: His Life and Work as Seen by His Friends and Colleagues*. Amsterdam: North-Holland, 1967.

Ruhla, Charles. *The Physics of Chance: From Blaise Pascal to Niels Bohr*. Oxford: Oxford University Press, 1992.

Rukeyser, Muriel. *Willard Gibbs*. Garden City, NY: Doubleday, Doran, & Co., 1942.

Russell, Bertrand. *The Autobiography of Bertrand Russell, 1914–1944, Volume II*. London: Allen & Unwin, 1967.

Sanders, John. *The God Who Risks: A Theology of Providence*. Downers Grove, IL: IVP, 1998.

Sartre, Jean-Paul. *Being and Nothingness: An Essay on Phenomenological Ontology*. New York: Philosophical Library, 1956.

Schrödinger, Erwin. *Collected Papers on Wave Mechanics: Together with His Four Lectures on Wave Mechanics*. New York: Chelsea, 1982.

———. *The Interpretation of Quantum Mechanics: Dublin Seminars (1949–1955) and Other Unpublished Essays*. Woodbridge, CT: Ox Bow, 1995.

Schweber, S. S. *QED and the Men Who Made It: Dyson, Feynman, Schwinger, and Tomonaga*. Princeton, NJ: Princeton University Press, 1994.

Scott, Drusilla. *Everyman Revisited: The Common Sense of Michael Polanyi*. Lewes, UK: Book Guild, 1985.

Segrè, Gino. *Faust in Copenhagen: A Struggle for the Soul of Physics*. New York: Viking, 2007.

Shaw, Luci. *Polishing the Petoskey Stone: New & Selected Poems*. Wheaton, IL: H. Shaw, 1990.

Shimony, Abner. *Search for a Naturalistic World View*. Cambridge: Cambridge University Press, 1993.

Simmons, Ernest L. *The Entangled Trinity: Quantum Physics and Theology*. Minneapolis: Fortress, 2014.

Sokal, Alan D. *Beyond the Hoax: Science, Philosophy, and Culture*. Oxford: Oxford University Press, 2008.

Sproul, R. C. *Not a Chance: The Myth of Chance in Modern Science and Cosmology*. Grand Rapids: Baker, 1994.

Stapp, Henry Pierce. "The Copenhagen Interpretation." *American Journal of Physics* 40.8 (1972) 1098–1116.

———. *Mind, Matter, and Quantum Mechanics*. Berlin: Springer-Verlag, 1993.

Stereogram builder. http://www.easystereogrambuilder.com/

Taylor, Edwin. "The Boundaries of Nature: Special and General Relativity and Quantum Mechanics, a Second Course in Physics." *American Journal of Physics* 66.5 (1998) 368–76.

Terrall, Mary. *The Man Who Flattened the Earth: Maupertuis and the Sciences in the Enlightenment*. Chicago: University of Chicago Press, 2002.

Tetrode H. "Über *den Wirkungszusammenhang der Welt der Klassische Dynamik*." *Zeitschrift für Physik* 10 (1922) 317–28.

Torrance, Thomas F. *Christian Theology and Scientific Culture*. New York: Oxford University Press, 1981.

———. *Divine and Contingent Order*. Oxford: Oxford University Press, 1981.

———. *Reality and Evangelical Theology*. Philadelphia: Westminster, 1982.

———. *Reality and Scientific Theology*. Edinburgh: Scottish Academic Press, 1985.
———. *Theological Science*. London: Oxford University Press, 1969.
———. *Transformation & Convergence in the Frame of Knowledge: Explorations in the Interrelations of Scientific and Theological Enterprise*. Grand Rapids: Eerdmans, 1984.
———. *Space, Time, and Incarnation*. Edinburgh: T. & T. Clark, 1997.
Torretti, Roberto. *The Philosophy of Physics*. Cambridge: Cambridge University, 1999.
Vanhoozer, Kevin J., James K. A. Smith, and Bruce Ellis Benson. *Hermeneutics at the Crossroads*. Bloomington, IN: Indiana University Press, 2006.
Vanhoozer, Kevin J., and Martin Warner. *Transcending Boundaries in Philosophy and Theology: Reason, Meaning and Experience*. Aldershot, UK: Ashgate, 2007.
Von Neumann, John, and Robert T. Beyer. *Mathematical Foundations of Quantum Mechanics*. Princeton, NJ: Princeton University Press, 1996.
———. *Mathematische Grundlagen der Quantenmechanik*. Berlin: Springer, 1932.
Wallace, David Foster. *This is Water: Some Thoughts, Delivered on a Significant Occasion about Living a Compassionate Life*. New York: Little, Brown, 2009.
Ward, Keith. *God, Chance & Necessity*. Oxford: One World, 1996.
Weizsäcker, Carl Friedrich. *The History of Nature*. Chicago: University of Chicago Press, 1949.
———. *The World View of Physics*. Chicago: University of Chicago Press, 1952.
Wheeler, J. A., and R. P. Feynman. "Interaction with the Absorber as the Mechanism of Radiation." *Reviews of Modern Physics* 17.2–3 (1945) 157–61.
Wheeler, John Archibald, and Wojciech Hubert Zurek. *Quantum Theory and Measure ment*. Princeton, NJ: Princeton University Press, 1983.
Whitaker, Andrew. *Einstein, Bohr, and the Quantum Dilemma*. Cambridge: Cambridge University Press, 1996.
Wigner, Eugene Paul. *Symmetries and Reflections; Scientific Essays of Eugene P. Wigner*. Bloomington, IN: Indiana University Press, 1967.
Wilson, Edward O. *Consilience: The Unity of Knowledge*. New York: Knopf, 1998.
Winkel, R. J., Grant R. Fowles, and George L. Cassiday. *Instructor's Solutions Manual to Accompany Analytical Mechanics*. 5th ed. by Fowles/Cassiday. Fort Worth, TX: Saunders, 1994.
Wolf, Fred Alan. *Taking the Quantum Leap: The New Physics for Nonscientists*. San Francisco: Harper & Row, 1981.
Zajonc, Arthur. *Catching the Light: The Entwined History of Light and Mind*. New York: Bantam, 1993.

Index

Page numbers in italics refer to a diagram, figure, table, or picture. The letter n after a page number indicates a footnote.

CPSIA information can be obtained
at www.ICGtesting.com
Printed in the USA
LVHW081500090620
657726LV00003B/464